35.00

APPLICATIONS OF ARTIFICIAL INTELLIGENCE FOR ORGANIC CHEMISTRY
The DENDRAL Project

McGraw-Hill Computer Science Series

Allen: *Anatomy of LISP*
Bell and Newell: *Computer Structures: Readings and Examples*
Donovan: *Systems Programming*
Feigenbaum and Feldman: *Computers and Thought*
Gear: *Computer Organization and Programming*
Givone: *Introduction to Switching Circuit Theory*
Goodman and Hedetniemi: *Introduction to the Design and Analysis of Algorithms*
Hamacher, Vranesic, and Zaky: *Computer Organization*
Hamming: *Introduction to Applied Numerical Analysis*
Hayes: *Computer Architecture and Organization*
Hellerman: *Digital Computer System Principles*
Hellerman and Conroy: *Computer System Performance*
Kain: *Automata Theory: Machines and Languages*
Katzan: *Microprogramming Primer*
Kohavi: *Switching and Finite Automate Theory*
Liu: *Elements of Discrete Mathematics*
Liu: *Introduction to Combinatorial Mathematics*
MacEwen: *Introduction to Computer Systems: Using the PDP-11 and Pascal*
Madnick and Donovan: *Operating Systems*
Manna: *Mathematical Theory of Computation*
Newman and Sproull: *Principles of Interactive Computer Graphics*
Nilsson: *Problem-Solving Methods in Artificial Intelligence*
Rosen: *Programming Systems and Languages*
Salton: *Automatic Information Organization and Retrieval*
Stone: *Introduction to Computer Organization and Data Structures*
Stone and Siewiorek: *Introduction to Computer Organization and Data Structures: PDP-11 Edition*
Tonge and Feldman: *Computing: An Introduction to Procedures and Procedure-Followers*
Tremblay and Bunt: *An Introduction to Computer Science: An Algorithmic Approach*
Tremblay and Manohar: *Discrete Mathematical Structures with Applications to Computer Science*
Tremblay and Sorenson: *An Introduction to Data Structures with Applications*
Tucker: *Programming Languages*
Watson: *Timesharing System Design Concepts*
Wiederhold: *Database Design*
Winston: *The Psychology of Computer Vision*

McGraw-Hill Advanced Computer Science Series

Davis and Lenat: *Knowledged-Based Systems in Artificial Intelligence*
Feigenbaum and Feldman: *Computers and Thought*
Kogge: *Pipelining and Overlap*
Lindsay, Buchanan, Feigenbaum, and Lederberg: *Applications of Artificial Intelligence for Organic Chemistry: The DENDRAL Project*
Nilsson: *Problem-Solving Methods in Artificial Intelligence*
Watson: *Timesharing System Design Concepts*
Winston: *The Psychology of Computer Vision*

APPLICATIONS OF ARTIFICIAL INTELLIGENCE FOR ORGANIC CHEMISTRY
The DENDRAL Project

Robert K. Lindsay
Research Scientist
University of Michigan

Bruce G. Buchanan
Adjunct Professor of Computer Science
Stanford University

Edward A. Feigenbaum
Professor of Computer Science
Stanford University

Joshua Lederberg
President, The Rockefeller University

Formerly, Professor of Genetics
Stanford University

McGraw-Hill Book Company

New York St. Louis San Francisco Auckland Beirut Bogotá Hamburg Johannesburg
Lisbon London Lucerne Madrid Mexico Montreal New Delhi Panama
Paris San Juan São Paulo Singapore Sydney Tokyo Toronto

APPLICATIONS OF ARTIFICIAL INTELLIGENCE
FOR ORGANIC CHEMISTRY
The DENDRAL Project

Copyright © 1980 by McGraw-Hill, Inc. All rights reserved.
Printed in the United States of America. No part of this publication
may be reproduced, stored in a retrieval system, or transmitted, in any
form or by and means, electronic, mechanical, photocopying, recording, or
otherwise, without the prior written permission of the publisher.

1 2 3 4 5 6 7 8 9 0 MAMA 8 9 8 7 6 5 4 3 2 1 0

This book was set in Times Roman by The Total Book/ACS.

Library of Congress Cataloging in Publication Data

Main entry under title:

Applications of artificial intelligence for organic
    chemistry.

    (McGraw-Hill advanced computer science series)
    Bibliography: p.
    Includes index.
    1. DENDRAL (Computer programs) 2. Artificial
intelligence—Data processing. 3. Chemistry, Organic—
Data processing. I. Lindsay, Robert K. II. Series.
QD255.5.E4A66   547'0028'5425   80-12508
ISBN 0-07-037895-9

CONTENTS

Foreword		ix
Preface		xi
1	**Introduction**	1
2	**The Structure Elucidation Problem of Organic Chemistry**	3
2.1	Introduction	3
2.2	Isomerism	4
2.3	Organic compounds and Nomenclature	5
2.4	Mass Spectrometry	13
2.5	Some Important Refinements of the MS Technique	19
2.6	Other Analytical Methods	24
2.7	Library Search	26
2.8	Summary	27
3	**Artificial Intelligence**	28
3.1	Introduction	28
3.2	Problem-Solving Methods	30
3.3	DENDRAL	35
3.4	Outline of DENDRAL Programs	39
4	**The DENDRAL Generator**	40
4.1	Introduction	40
4.2	Overview	41
4.3	Ring Generation	44
4.4	Tree Generation—The Acyclic Generator	48
4.5	The Cyclic Generator	53
4.6	Constraining the Generator: CONGEN	53

5 Heuristic DENDRAL Planning — 68

5.1 Introduction — 68
5.2 The Early Planner and the Planning Rule Generator — 69
5.3 MOLION — 70
5.4 Empirical Formula — 78
5.5 Generalized Break Analysis — 78
5.6 Conclusion — 85

6 Heuristic DENDRAL Testing — 86

6.1 Introduction — 86
6.2 Predictor Production System — 87
6.3 Graph Structure and Production Representation — 88
6.4 Ranking the Candidate Explanations — 100
6.5 Summary of Heuristic DENDRAL — 105

7 Meta-DENDRAL — 107

7.1 Introduction — 107
7.2 INTSUM—Data Interpretation and Summary — 109
7.3 RULEGENeration — 115
7.4 RULEMODification — 123
7.5 Summary — 125

8 Results — 126

8.1 Introduction — 126
8.2 The Scope of Structural Isomerism — 127
8.3 Acyclic Heuristic DENDRAL — 130
8.4 CONGEN Results — 134
8.5 Planner Results — 135
8.6 Meta-DENDRAL Results — 143
8.7 DENDRAL Predictor Results — 144
8.8 Design Principles — 147

9 Summary and Conclusions — 153

9.1 Introduction — 153
9.2 Knowledge Engineering — 153
9.3 Scientific Discovery — 160
9.4 The Prospects for Automatic Science — 168

10 Project Publications — 169

References — 179
Name Index — 187
Subject Index — 189

FOREWORD

It is no pure coincidence that an organic chemist rather than a computer specialist is writing this foreword. In my opinion the most significant aspect of the development of DENDRAL is that it can be appreciated and used by a community of scholars that encompasses a diverse range from mathematics and computer sciences all the way to chemistry. The former group has been attracted by DENDRAL's intrinsic interest and the latter primarily by its applications. As an organic chemist let me emphasize the applications.

The use of computers as a computational and data acquisition tool is accepted by most chemists. This use is especially true as an adjunct to instrumentation (for instance NMR spectrometers, automated x-ray diffractometers, etc.) and to library searches or spectral file examinations. The use of computers in the manipulation of symbolic rather than numerical inputs is of much more recent origin and until recently has been ignored, and for psychologically understandable reasons even opposed, by organic chemists. I emphasize the organic chemical community because together with biochemistry, it encompasses well over half of all practicing chemists and involves the least amount of sophisticated mathematics.

Symbolic manipulations by computers are in principle important in two areas of chemistry—synthesis and structure elucidation. It is the former where the use of computers has not been widely accepted because of the fear that thinking man will simply be reduced to an appendage to a machine. The synthetic chemist wishes to be both architect and building contractor—the former function being the intellectually and aesthetically more pleasing one—and it is precisely this architectural role that the computer is perceived partially to usurp.

The structural chemist, on the other hand, has always been receptive to aid from many different areas—notably a variety of instrumental methods; indeed most physical methods have entered general organic chemical methodology through the structural chemist's interests and efforts. It is not surprising, therefore, that computer-aided structure elucidation has found more favor than computer-aided design of organic synthesis. While this argument is primarily emotional, there is also a logical one. No synthetic organic chemist claims, or needs to claim, that he or she has thought of all possible synthetic paths to a given molecule. The structural chemist, on the other hand, must be able to claim that every possible structure compatible with the available chemical and physical evidence has been considered. It is here that computer-aided structure elucidation plays an extraordinarily important role, and it is here that DENDRAL has made it possible for enormous advances to have occurred in a period of less than 10 years.

It is no mere coincidence that scientists from as diverse disciplines as genetics, computer sciences, and chemistry collaborated in developing DENDRAL as a logical and practical concept. The numerous applications of DENDRAL and related computer programs have now been documented in many scientific publications, and putting all this material, including the historical background, into one single volume constitutes an important service for scientists from many fields.

Carl Djerassi
Professor of Chemistry
Stanford University

PREFACE

Some explanation of the history and authorship of this book is in order. I began writing this book while on sabbatical leave from the University of Michigan in 1975. I had chosen to spend that year at Stanford University to learn firsthand about the DENDRAL Project. My interest derived from a continuing desire to understand what has been accomplished by the field of artificial intelligence, with which I have long been associated but toward which I have tried to maintain a critical stance. DENDRAL is widely claimed to be one of the most notable successes of this field. I wondered what generalizable lessons it had to share.

Ed Feigenbaum suggested that I put my efforts, and my perspective as a sympathetic but critical Project outsider, to productive use by writing a volume summarizing the DENDRAL research, bringing together in one place for the student, and for archival purposes as well, the threads of work that had been strung here and there throughout the literature of computer science and analytic organic chemistry. I agreed to do so.

This background partially explains the pedigree of this volume. The list of authors might have included every one of the many contributors to the research, but it has been limited to the major originators and long-term directors of the computer science directions of the Project, plus myself. As it has developed, I have written almost all the text; Bruce Buchanan has made major contributions of text and reviewed every draft of the entire manuscript. Edward Feigenbaum and Joshua Lederberg, of course, have been the major forces directing the entire project and gave invaluable assistance and consultation in the preparation of this book.

A large number of people remain who deserve a lot of credit. I would first like to thank Harold Brown, Ray Carhart, Geoff Dromey, and Dennis Smith, who directly helped me to understand specific aspects of the Project, and who reviewed my telling of those stories. Each of them has devoted a generous amount of time to this effort. Also, I would like to give special acknowledgment and thanks to Maija Kibens for her help with the entire work, and to Nils Nilsson for his valuable comments.

Additionally, there are many contributors to the Project itself who deserve acknowledgment in this volume. Those, in addition to the authors, who are responsible for most of the artificial intelligence concepts are Raymond Carhart, Carl Djerassi, Dennis Smith, Harold Brown, Allan Delfino, Geoff Dromey, Alan Duffield, Neil Gray, Larry Masinter, Tom Mitchell, James Nourse, N. S. Sridharan, Georgia Sutherland, and William White.

Other Project contributors have been M. Achenbach, C. Van Antwerp, A. Buchs, L. Creary, L. Dunham, H. Eggert, R. Engelmore, F. Fisher, R. Gritter, S. Hammerum, L. Hjelmeland, A. Lavanchy, S. Johnson, J. Konopelski, K. Morrill, T. Rindfleisch, A. Robertson, G. Schroll, G. Schwenzer, Y. Sheikh, M. Stefik, T. Varkony, A. Wegmann, W. Yeager, and A. Yeo.

The financial sponsorship of such an extended effort is extremely important. The Project has been made possible by the vision of funding-agency executives who have realized the importance of long-range commitments for such research. In its early years DENDRAL research was sponsored by the National Aeronautics and Space Administration and the Advanced Research Projects Agency of the Department of Defense. More recently the Project has been sponsored by the National Institutes of Health (Grant RR-00612). The Project depends on the SUMEX computing facility located at Stanford University. This facility is sponsored by the National Institutes of Health (Grant RR-00785) as a national resource for applications of artificial intelligence to medicine and biology.

The expository portions of the manuscript have been written for some time, but the completion of the book has been delayed. The main reason for this delay, aside from the inevitable desire to include mention of each new development in a project that will never be completed, is in the difficulty we, the authors, have had in summarizing (and agreeing on) what we see to be the Project's lessons for computer science and artificial intelligence. Its appearance at this time does not signal the resolution of our problem as much as our frustration with it and with the time it has taken us to formulate our answers. Needless to say, we are not entirely happy with the result. Hopes and visions, we suppose, always seem much more grand than can be forcefully argued to others. We have tried not to overstate the case; I hope the reader will see some of the vision nonetheless. I especially hope that the importance of some of the specific insights the project members have developed will be appreciated by and will benefit at least the sympathetic readers.

This book is written primarily for an audience of computer scientists (not just artificial intelligence researchers), but it will be comprehensible to most nonspecialists who have a general technical background. The book presumes no knowledge of chemistry beyond the level of an introductory college course on general chemistry and no knowledge of computer science beyond the level of a basic course in computer programming. Our aim is to describe and evaluate the Project's work as an example of artificial intelligence research, and only secondarily to discuss its importance to chemistry.

Robert Lindsay
Ann Arbor, Michigan

APPLICATIONS OF ARTIFICIAL INTELLIGENCE FOR ORGANIC CHEMISTRY
The DENDRAL Project

CHAPTER
ONE

INTRODUCTION

The DENDRAL Project began with Lederberg's construction in 1964 of an algorithm for generating canonical names and structural descriptions of molecules. In 1965 the Project's goals broadened to include interpretation of analytical chemical data using methods of artificial intelligence.

The DENDRAL Project is a study of scientific reasoning. More specifically, it is an application of computer science to the problem of molecular *structure elucidation* in organic chemistry: the determination of the topological structure of organic compounds from indirect observations of these compounds with the empirical procedures of modern chemistry such as mass spectrometry. The computer programs that are the result of this work are products of *artificial intelligence* (AI) research, the branch of computer science that undertakes the challenging but controversial task of mechanizing perception and thought. AI is distinguished from other applications of computers by its attention to problems for which no straightforward, assured solution methods are known in advance. In particular, the programs we will discuss employ guessing strategies and similar rules of thumb called *heuristics*. This approach to artificial intelligence is called *heuristic programming*. Our book describes the structure elucidation problem, the DENDRAL programs, and the current directions of the Project.

Within computer science the DENDRAL Project is noteworthy in several ways. It was the first major application of heuristic programming to experimental analysis in an empirical science, a practical problem of some importance. It was the first large-scale program to embody the strategy of using detailed, task-specific knowledge about the problem domain as a source of heuristics, and to seek generality through automating the acquisition of such knowledge. It has achieved a high level of performance because it uses a substantial amount of knowledge of chemistry. It is one of the larger, more sustained AI projects undertaken, giving it a certain prominence even apart from its successes. It is being used by chemists, other than its developers, in the pursuit of their own research goals. It is an interdisciplinary project that has been continuously productive for over a decade. Perhaps most significant is that this research is an extensive empirical exploration of heuristic programming techniques; as such it is a validation of the strengths and weaknesses of these techniques and an instantiation of a philosophical concept of automatic discovery procedures whose status has long been in dispute.

"DENDRAL" is the name of the project and also the name of the programs, sometimes further distinguished as *Heuristic* DENDRAL and *Meta*-DENDRAL. DENDRAL originally stood for DENDRitic ALgorithm, a procedure for exhaustively and nonredundantly enumerating all the topologically distinct arrangements of any given set of atoms, consistent with the rules of chemical valence. The original algorithm, which generated only ring-free (acyclic) structures (i.e., the aliphatic compounds) was devised by one of the authors (Joshua Lederberg). The dendritic algorithm (as well as its later version, which in addition encompasses ring structures) is the heart of the DENDRAL programs. This algorithm defines the set of *possible* solutions through which the DENDRAL programs search for *likely* solutions. A basic feature of DENDRAL, and an important limitation on the range of applicability of its methods, is its *uniform notation for hypotheses*, here taking the form of graphs in the abstract sense of points (nodes) linked by lines (edges), depicting, respectively, the atoms and bonds of molecules.

The means by which these programs reduce the set of possible chemical graphs (solutions to the elucidation problem) is the real story of DENDRAL. The heuristics employed are based on judgment and specific chemical knowledge, the kinds of expertise that are popularly called intuition. Heuristic DENDRAL comprises the programs that employ these methods. An important point, one on which the success of the project has turned, is that the constructors of the DENDRAL programs eschewed the search for general principles of problem solving or learning in favor of specific knowledge of a special problem. The success of this project against the general background of failure of the search for general systems does not, of course, decide the issue of better approach; but it ought to attract careful attention.

The foregoing does not deny the existence of general discovery principles for science. Rather, generality, if the DENDRAL approach is correct, is to be found at a more abstract level. That is, there may exist general principles for acquiring specific knowledge, and general principles for selecting specific methods of applying that knowledge. One attempt to program more general rule-discovering (learning) methods is called Meta-DENDRAL, to distinguish it from the original (performance) system.

Currently much of the Project's effort is devoted to the CONGEN program. This program, the CONstrained GENerator, embodies the general (acyclic and cyclic) generation algorithm in a system that allows the chemist to constrain its enumeration in a variety of ways. CONGEN is being made available to interested nonproject chemists, and every effort is being made to provide assistance in its application so that it will be as readily accessible as is possible for a system of this complexity.

The DENDRAL project, extensive as it is, has been the subject of much attention. Over 80 scientific papers and a number of popular descriptions have been published, and the project has received a measure of attention in the review literature. This book, however, is the first attempt to provide a comprehensive description of the goals and accomplishments of the entire decade and a half of research. We hope it will provide a definitive, self-contained source of information about this work.

CHAPTER
TWO

THE STRUCTURE ELUCIDATION PROBLEM OF ORGANIC CHEMISTRY

The DENDRAL programs are designed to aid organic chemists interpret data from samples of unknown compounds. Even after the numbers and types of atoms of a compound are determined, the problem remains of deciding how the atoms are connected in the molecule. Data useful in structure determination may come from a variety of analytic instruments, as well as from chemical experiments. A mass spectrometer is one such instrument that provides valuable structural information.

2.1 INTRODUCTION

This chapter describes the chemistry problems addressed by the DENDRAL programs, including a description of the technique of mass spectrometry which is a major source of data from which the system works. Also included is a brief introduction to some of the terminology and concepts of organic chemistry. The intention is to make this book self-contained by gathering here the necessary definitions and facts of organic chemistry needed to understand the remaining chapters. An extensive index is provided so that this chapter may be used as a reference.

The general problem to which the DENDRAL programs apply is an important, substantive problem in organic chemistry: structure elucidation, that is, the determination of the organization of sets of atoms in specific molecules. The problem is important because the chemical and physical properties of compounds are determined not just by their constituent atoms, but by the arrangement of these atoms as well. Several empirical means are available for obtaining information about the structure of a compound. Prominent among these is mass spectrometry, and DENDRAL originally addressed problems associated only with this method. DENDRAL has since evolved to deal with the problems of structure elucidation on more general terms. In order to understand the problem that DENDRAL faces, it is necessary to introduce some elementary facts about organic chemistry and mass spectrometry.

2.2 ISOMERISM

Several models of molecules are used in chemistry. The one appropriate to our purposes is the ball-and-stick model, in which the "balls" are atoms, and the "sticks" are bonds between atoms. The bonds may be of different types, corresponding to the number of electrons shared by two atoms to form the bonds. Single, double, and triple bonds, denoted "—", "=", and "≡", respectively, correspond to two, four, and six shared electrons. A ball-and-stick model of the water molecule is presented in Figure 2-1.

It is known that a water molecule is composed of two hydrogen atoms and one oxygen atom. This information about the molecular structure is given by the *empirical formula* of the molecule: H_2O in the case of water. A given collection of atoms may, of course, be arranged in numerous ways. If atoms were really just balls of various sizes that could be stacked up in arbitrary arrangements, there would be many possible configurations. This number is limited by the known facts of chemical *valence*, i.e., the number of bonding sites available to an atom. These are *topological* constraints on the set of all possible organizations of atoms. Other constraints are *geometric*; these limit the molecular possibilities in accordance with facts about bond lengths and the angles between bonds. In our example, it is known that the topology of water is H—O—H, not H—H—O, because hydrogen has a valence of 1 not 2, and oxygen has a valence of 2 not 1. It is known further that the geometry of the water molecule, as shown in Figure 2-1, is such that the lengths of the bonds (the distances between atomic centers) is 0.0965 millimicron (a millimicron is 10^{-9} meter), and that the angle between the bonds is 104.5°.

It is useful at times to ignore some of the known constraints on molecular organization. Molecular organizations that are the same up to a point, but differ in some further regard, are called *isomers*. For example, two molecules with the same empirical formula but different topology (connectivity) are called structural isomers or *connectivity isomers*. We will use the latter term since it is more explicit. Usually the term "isomers," without further qualification, denotes connectivity isomers. Two molecules that have the same empirical formula and the same connectivity, but are not congruent (superimposable) in three-dimensional space, are *stereoisomers*. One type of stereoisomer arises because double bonds cannot in fact undergo the topology-preserving

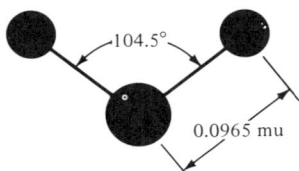

Figure 2-1 A water molecule. DENDRAL does not consider the lengths of bonds or the angles between them.

transformation of twisting. Thus these two arrangements

$$\begin{array}{c}H\\CH_3\end{array}\!\!>\!\!C\!\!=\!\!O\!\!<\!\!\begin{array}{c}H\\CH_3\end{array} \qquad \begin{array}{c}CH_3\\H\end{array}\!\!>\!\!C\!\!=\!\!O\!\!<\!\!\begin{array}{c}H\\CH_3\end{array}$$

are topologically equivalent but chemically distinct. In general, different isomers have different chemical and physical properties.

DENDRAL is mostly concerned with connectivity information about molecules because the context of its development was mass spectrometry, which is largely insensitive to stereoisomerism. Thus to most of DENDRAL, different geometric forms of the same topology are entirely equivalent.[1] The identification of a molecule means at least that its topological organization is known. One representation of molecular topology is the chemical graph.[2] In such graphs atoms are represented by nodes and bonds by edges connecting the nodes.

This graphical representation of chemical structure is used to considerable advantage by DENDRAL. DENDRAL's immediate and final hypotheses and its problem definition are all characterized by the common language of graphs. Thus there is a single representation language in which the various parts of DENDRAL communicate.

Graphs are often abbreviated when the nature of the bonding is invariant for certain substituents. Thus the notation CF_3-CN looks like an empirical formula but is actually a graph in abbreviated form, it being understood that the three fluorine atoms (F's) adjacent to the left-hand carbon atom (C) are singly bonded to the carbon atom, which in turn is singly bonded to the carbon in the right constituent; and that the carbon and nitrogen (N) in the right constituent are triply bonded. When bonds are not indicated explicitly, i.e., two atom names are simply juxtaposed, it is necessary to determine the nature of the implied bond (single, double, or triple) by resorting to knowledge of valences. For example, in R_1-CO-R_2, with the R's being abbreviations for constituents singly bonded to the carbon,[3] it is understood that R_2 is bonded to C and the carbon to oxygen bond is double, since the valence of oxygen is 2. The most frequent abbreviation is the omission of hydrogen atoms. All bonding sites that are unspecified are thus assumed to be bonded to hydrogen atoms. We will adopt this convention throughout. In many illustrations, carbon atoms are left as unnamed nodes of graphs.

2.3 ORGANIC COMPOUNDS AND NOMENCLATURE

In this book reference will be made to several classes of compounds. Definitions of these are gathered here for ease of reference. This section may be skipped by readers familiar with the basic terminology of organic chemistry.

[1] Recent work by James Nourse, described in Sec. 4.6.3, is an exception to this statement.
[2] Unfortunately, *graph* commonly refers to a plot of one variable as a function of another. In this book, *graph* will be used exclusively in the sense of mathematical graph theory: a set of discrete points, called nodes, connected by lines, called edges.
[3] In keeping with conventional chemical notation, R will denote arbitrary substructures of molecules.

An element of valence 4 offers more possibilities for forming compounds than does any other. Carbon is the lightest and most abundant element of this valence, and thus the chemistry of carbon compounds, organic chemistry, comprises a rich and enormous field, including most biologically active compounds.

Hydrocarbons are compounds composed solely of hydrogen and carbon. *Saturated* hydrocarbons are those in which all carbon-carbon bonds are single, and the remaining bonding potential of the carbon atoms is used by hydrogen atoms. These compounds are saturated in the sense that they contain as much hydrogen as possible.

The saturated hydrocarbons are also called *alkanes*, the simplest of which is methane (CH_4). In methane each hydrogen is singly bonded to the single carbon atom. A chain of two carbon atoms fully saturated (i.e., six hydrogens) is called ethane; three carbons fully saturated is propane. The general empirical formula for alkanes is $C_n H_{2n+2}$. Some examples are diagrammed in Figure 2-2.

If a hydrocarbon of n carbons contains fewer than $2n + 2$ hydrogens, it is *unsaturated*. This state will occur only if two or more carbons are multiply bonded to one another or linked into a ring. Each double bond will reduce the hydrogen count by 2, each triple bond by 4. The amount by which a hydrocarbon falls short of saturation is measured by the *degree of unsaturation*, which is defined as half the difference between $2n + 2$ and the actual hydrogen count. This concept is explored more fully in Section 2.3.3.

A *radical* or *group* is a substructure of a molecule that does not exist as a stable compound because it has bonding potential (valence). If one hydrogen is missing from a methane molecule, we are left with the radical —CH_3, called *methyl*. Similarly, removing a hydrogen from the other alkanes results in a radical whose name is the same as that of its parent alkane, but with the suffix "-yl" in place of "-ane": ethyl, propyl, etc. These radicals are referred to generically as *alkyls* (Figure 2-2).

We have discussed only those alkanes in which the carbons line up in a chain. If there are more than three carbons, however, they might be arranged in a branching structure. The last alkane in Figure 2-2 is a branched version. It contains four carbons, but is not butane because they are not chained. Rather it is given a compound name deriving from the name of the molecule associated with its longest chain, preceded by the names of the radicals that form the side chains. The compound depicted is methylpropane. Butane and methylpropane have the same empirical formula (C_4H_{10}), but different structures. That is, they are connectivity isomers and have different chemical properties.

For longer main chains, not only might different radicals be side chains, but the same radical might be in a different location. This phenomenon is not possible in the case of methylpropane, because putting the methyl radical on the end would yield butane. The naming conventions become increasingly complex and are not important for our purposes.[4]

Alkenes are ring-free hydrocarbons, but are unsaturated because there is double bonding between at least one pair of carbons and thus fewer hydrogens than in an

---

[4]One of the original motivations for developing the DENDRAL notation (a linear canonical naming convention described in Section 4.4.1) was the confusion resulting from the traditional naming schemes.

alkane containing the same number of carbons. *Alkynes* are ring-free unsaturated hydrocarbons containing one or more carbon-carbon triple bonds. The locations of the multiple bonds in the carbon chain must also be accounted for in naming and picturing the alkenes and alkynes.

The *aromatic hydrocarbons* are compounds containing one or more aromatic rings, an important structure illustrated in Figure 2-3. It is not possible to represent this structure correctly in such a diagram, because the bonds are *equivalent*; they are not single and double bonds, as suggested by the figure. Rather all bonds are something "in between" single and double. They are said to resonate, but to explain further involves detailed understanding of the role of orbital electrons in chemical bonding, a subject that is involved and not important for our purposes.

Other important classes of organic compounds may be considered as derivatives of the hydrocarbons. They contain *heteroatoms*—atoms other than carbon and hydrogen. Certain groups containing heteroatoms are called *functional groups*, and are important in the classification of compounds because their presence usually is predictive of one or more significant chemical and physical properties. They thus define families of similar compounds.

Alcohols result when a *hydroxyl* radical (—OH) replaces a hydrogen. The simplest alcohol is *methanol*, CH_3—OH. Although extrapolation suggests that the simplest "alcohol" is water, H—OH, it is not classified as an alcohol because it does not share the physical and chemical properties of this class. The *ethers* are closely related to the alcohols. The simplest ether derives from methanol and is CH_3—O—CH_3, dimethyl ether. In general, replacing (substituting) the H of an OH radical in an alcohol with an alkyl radical results in an ether. The names of alcohols and ethers derive from the name of the chain containing the defining radical. Further examples appear in Figure 2-4.

A *carbonyl* group is a carbon doubly bonded to an oxygen:

$$-\overset{|}{C}=O$$

If the carbonyl is bonded to two hydrogens *or* one hydrogen and one alkyl, it forms an *aldehyde*. If it is bonded to *two* alkyl groups, it is a *keto group*, and the compound is a *ketone*. See Figure 2-5.

Oxidation of the aldehyde radical

$$-\overset{H}{\underset{|}{C}}=O$$

yields the *carboxyl* radical

$$-\overset{OH}{\underset{|}{C}}=O$$

Compounds containing this radical are *carboxylic acids*. The simplest of these are methanoic acid, H—CO—OH, and acetic acid, CH_3—CO—OH. These react with alcohols to produce *esters*, of which fats and oils are important subclasses.

8 APPLICATIONS OF ARTIFICIAL INTELLIGENCE FOR ORGANIC CHEMISTRY

Amines and *amides* contain *nitrogen*. They may be considered derivatives of the basic nitrogen compound NH_3, ammonia. Amines result when one or more of the hydrogens is replaced by an alkyl group. Amides result from a reaction between ammonia and carboxylic acids. See Figure 2-6.

An important subclass of carboxylic acids is that which contains an amino group, $-NH_2$, on a carbon adjacent to the carbon of a carboxyl group. These compounds are the *amino acids*, from which all proteins are formed.

The *thiols* are *sulfur* compounds that are analogs of the alcohols, the sulfhydryl group $-SH$ playing the role played by $-OH$ in alcohols. Derivatives of thiols are analogous to the derivatives of alcohols: *thioethers* and *thioesters*. They appear in proteins and since they are easily oxidized they frequently convert to $-S-S-$ links that attach two proteins, or a single protein to itself. These linkages, among others, are im-

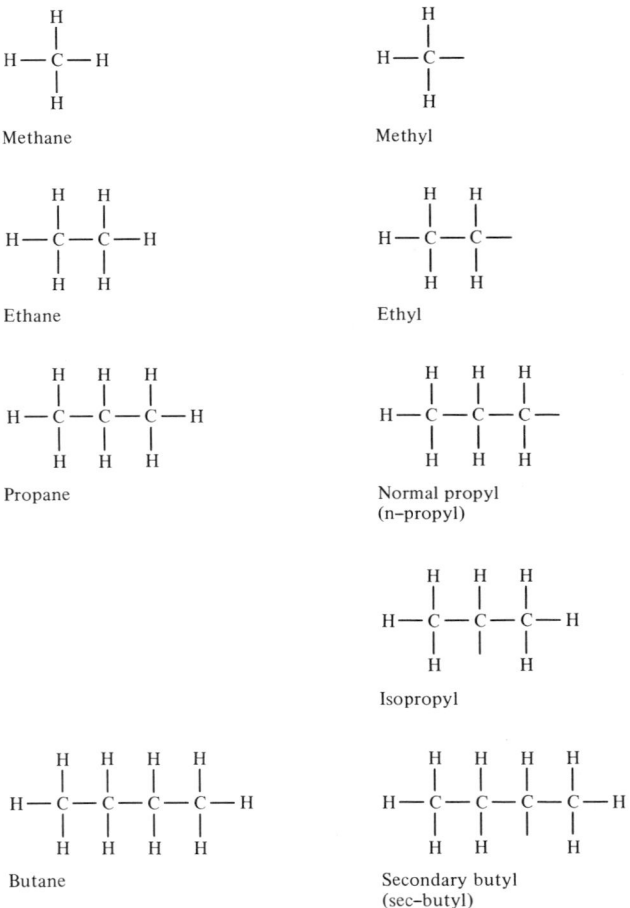

Figure 2-2 Alkanes and alkyl radicals.

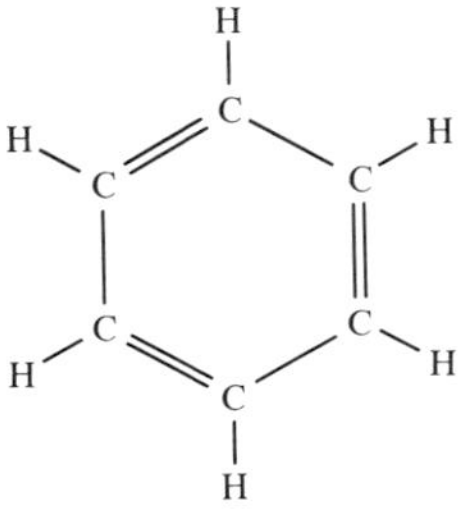

Figure 2-3 Benzene (a single aromatic ring).

portant in determining the three-dimensional geometry of protein molecules, from which many important biological properties derive.

2.3.1 Aliphatic versus Cyclic Compounds

The examples given above, with the exception of the aromatic compounds, have been described as structures with no rings. Such structures are called *trees* in graph theory, and *aliphatic* or *acyclic* compounds in chemistry. They are characterized by the fact that cutting any bond necessarily results in two separate pieces. All the classes discussed, except the alkanes, also contain members with rings. For example, consider the cyclic ketone depicted in Figure 2-7. In addition, there are important classes of compounds defined by particular ring structures. The *estrogens* are an example. Estrogens are members of the class of compounds called *steroids* characterized by a kernel structure consisting of four connected rings, as seen in Figure 2-8. *Sterols* are another type of steroid based on the same kernel structure.

2.3.2 First DENDRAL Applications

The first versions of DENDRAL were applied to the mass spectra of aliphatic compounds only, since an enumeration algorithm for cyclic compounds had not yet been

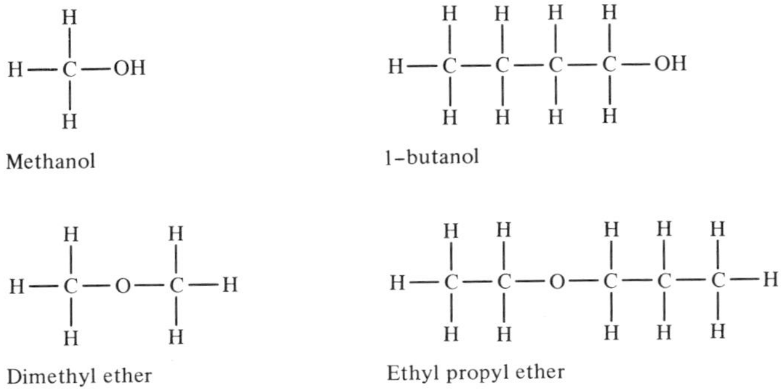

Methanol

1-butanol

Dimethyl ether

Ethyl propyl ether

Figure 2-4 Alcohols and ethers.

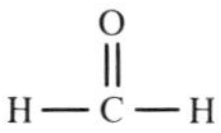

Formaldehyde
(methanal)

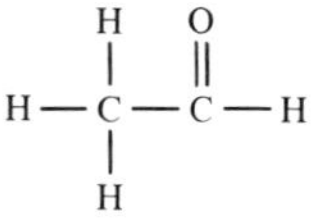

Acetaldehyde
(ethanal)

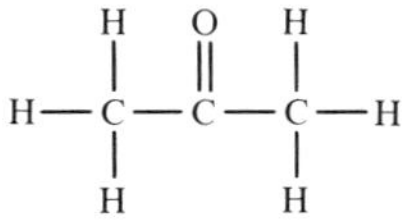

Acetone
(propanone)

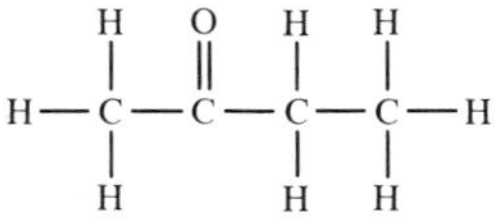

Methyl ethyl ketone
(butanone)

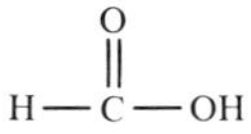

Methanoic acid
(formic acid)

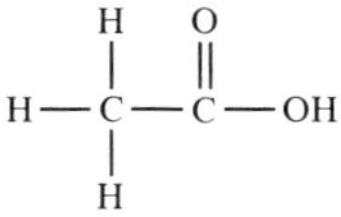

Acetic acid

Figure 2-5 Aldehydes and ketones.

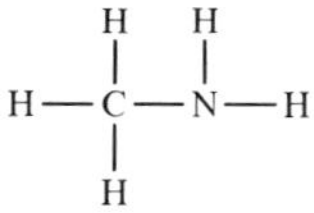

Methyl amine

Acetic acid + Ammonia → Acetamide + Water

Figure 2-6 Amines and amides.

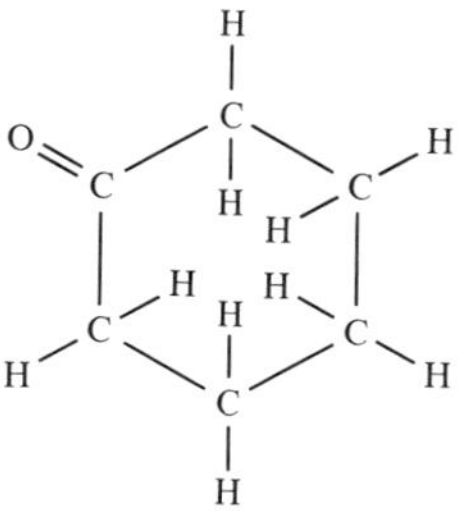

Figure 2-7 A cyclic ketone (cyclohexanone).

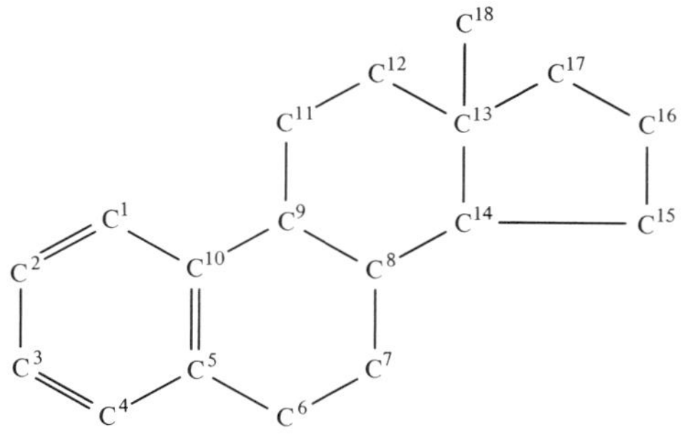

Figure 2-8 The estrogen skeleton. Numbers associated with the carbon atoms reflect the conventional numbering of node positions.

programmed. The compounds studied were, in roughly chronological order, amino acids, ketones, ethers, alcohols, amines, thiols, and thioethers.

Later versions of DENDRAL, incorporating a version of the cyclic structure generator, have been applied primarily to steroids, in particular estrogens, marine sterols, and related compounds.

The DENDRAL programs are not limited in application to these classes of compounds, but are general mechanisms that could be applied to any compounds for which certain types of information are available. A practical limit on size of molecules amenable to the DENDRAL methods is, roughly, 100 atoms. This number is approximately the limit of mass spectrometric methods as well. Thus proteins, which are generally much larger, are among the compounds to which the DENDRAL system (and mass spectrometry) do not apply.

As will be described in Chapter 8, the above list of applications was selected in part for their value in developing the DENDRAL concepts and in part because they were of interest for their importance to contemporary chemistry.

2.3.3 Degree of Unsaturation

It may surprise those who are not chemists to learn that, given an empirical formula, it is possible to determine exactly the number of cycles (rings and multiple bonds) each connectivity isomer of that formula must contain. This number is the degree of unsaturation of the empirical formula, a generalization of the concept introduced for hydrocarbons.

We note that univalent atoms, such as H, Cl, and F, cannot be part of a ring, since each atom of a ring must be connected to at least two other atoms, otherwise there would be no closure. Recall that a saturated hydrocarbon is one that contains the maximum possible number of hydrogens, and hence can contain no carbon-carbon double or triple bonds. It is also true that a hydrocarbon containing a ring cannot be saturated, since two of the carbon valences are used not for hydrogen atoms but to "close the ring."

Two doubly bonded atoms may be considered to be a small ring, and two atoms triply bonded may be considered as two small, connected rings. In fact, DENDRAL so considers them, treating them in the same way as any other ring structures, even though these bonds are not normally thought of in this way in chemistry.

Indeed, there is some ambiguity in the concept of a ring. For example, two edge-fused rings

may be viewed as two rings, or as three rings if we count the outer perimeter as a distinct ring. DENDRAL counts such structures as two rings, although some chemists count it as three rings.[5] DENDRAL treats any triple bond as two rings, thereby being at variance with some chemists. The virtue of DENDRAL's convention, aside from topological consistency, is that the computation of degree of unsaturation, defined below, is straightforward. *Hereafter, "ring" and "rings" refer to double and triple bonds as well as larger ring structures.* It follows that a saturated hydrocarbon is one that contains no rings and conversely that any molecules with rings must be unsaturated. In fact the degree of unsaturation of a set of atoms is defined as the number of rings that must be present in any molecules formed from the set. A molecule containing n rings will have a degree of unsaturation of n. For example, benzene (Figure 2-3) has a degree of unsaturation of 4.

The concept of degree of unsaturation generalizes to molecules other than hydrocarbons: the degree of unsaturation of a compound is the difference between the maximum possible number of univalent atoms the set of remaining atoms *could* bond and the actual number of univalent atoms available; it equals the number of rings in the compound's structure. It is important to understand this notion, so we will examine some examples. Consider the empirical formula $C_{10}H_{19}N$. Suppose these atoms were combined into a connected, ring-free structure. The sum of the valences of the nonunivalent atoms (10 carbons at 4 each plus 1 nitrogen at 3) is 43. Of these valences, the nonunivalent atoms (11 of them) will use two fewer than twice their number $(22 - 2 = 20)$ just bonding to one another. (To account for the reduction by two, consider the special case where they are connected into a single chain: each atom uses two bonds to connect to its immediate neighbors except for the two end atoms that use only one bond each. The number of bonds required to maintain connectedness remains the same if branching occurs.) Thus 23 valences remain to be divided between the univalent atoms and the formation of rings. Since there are 19 hydrogens, we have 4 valences remaining for the formation of rings; each ring closure requires 2 valences, thus we have *two unsaturations*. The empirical formula may be rewritten as $C_{10}N_1U_2$ to indicate this fact. All molecules with this formula will contain 2 rings and have 19 free valences. It is understood that all free valences will be occupied by H's, which by convention are not indicated.

To determine the degree of unsaturation from an empirical formula, compute

$$U = \frac{[S - (2K - 2) - F]}{2} \qquad \text{(Eq. 1)}$$

[5] However, it is not conventional to consider the triple bond, a topologically equivalent case, as three rings or, indeed, to consider it a ring structure at all.

where S = sum of valences of multivalent atoms
 K = the number of multivalent atoms
 F = number of free valences (univalent atoms)
 U = degree of unsaturation

Here are more examples of the calculation.

Empirical formula	Abbreviated formula
C_6H_9N	C_6NU_3
$C_{10}H_{19}N$	$C_{10}NU_2$
C_2F_3N	$C_2F_3NU_2$
C_6H_6 (e.g., benzene)	C_6U_4
$C_{18}H_{24}O_2$ (e.g., estradiol)	$C_{18}O_2U_7$

2.4 MASS SPECTROMETRY

2.4.1 The Instrument

Atoms and bonds are, of course, much too small to be examined by optical means. To determine the structure of a molecule of any complexity, chemists must resort to indirect methods. One of the most fruitful of these is mass spectrometry (MS). The essence of the technique is to break large molecules into fragments, infer the composition of these fragments, and use this information to guess the structure of a molecule that would break in the observed manner. The mass spectrometer breaks molecules into fragments by bombarding them with electrons at high energy, causing bonds to cleave. A molecule does not always end up in just two fragments; some fragments are broken further. Of course it is not possible to examine only one molecule at a time, and this fact complicates the picture. An electron beam breaks many molecules of a sample, and these do not all fragment in the same places, although some bonds are more subject to breaking than others. For example, single bonds generally break more readily than double bonds.

Also, certain groups are relatively immune from breaking in the mass spectrometer. For example, C=O will generally break off as a unit or remain as a constituent of a larger unit, but will seldom itself break.[6]

Determining the masses of the fragments requires that they be sorted. To accomplish this process, they are first accelerated. Since most of the fragments are electrically charged (that is, they are ions, normally singly charged), they can be accelerated by an electric field. Only positively charged ions are examined, since these are normally much more abundant than negative ions. The beam of ions thus produced passes into a magnetic field perpendicular to its path and is deflected, in accordance with the laws of electromagnetism and mechanics.

---

[6]The bonds connecting such a group to a structure are said to be alpha to the group. Thus R—CO—R has two bonds alpha to the C=O group. Similarly, bonds more distant from a group are called beta, gamma, and so forth. Breakage of an alpha bond is called *alpha cleavage*; similarly we have *beta cleavage, gamma cleavage*, and so forth.

Figure 2-9 A mass spectrometer with friend (Dr. Raymond Carhart). *(Photograph by Robert Lindsay.)*

Note that, from Newton's second law of motion, acceleration a equals force/mass; that the time t that an ion spends being deflected equals the length of the deflecting section divided by the ion's velocity v in that direction; and that the deflection is the distance traveled due to the application of a force to a particle with zero initial velocity in the direction of the force (i.e., perpendicular to its initial velocity), and thus is $\frac{1}{2} at^2$. It follows that the deflection is proportional to the force and inversely proportional to mv^2.

The *force* exerted on a moving ion in the magnetic field is directly proportional to its charge e and its velocity; therefore, the amount by which an ion is *deflected* from a straight path by the magnetic field is directly proportional to its charge and inversely proportional to its mass and its velocity. The result is that ions of the same mass-to-charge ratio m/e entering the magnetic section with the same velocity are deflected by the same amount.

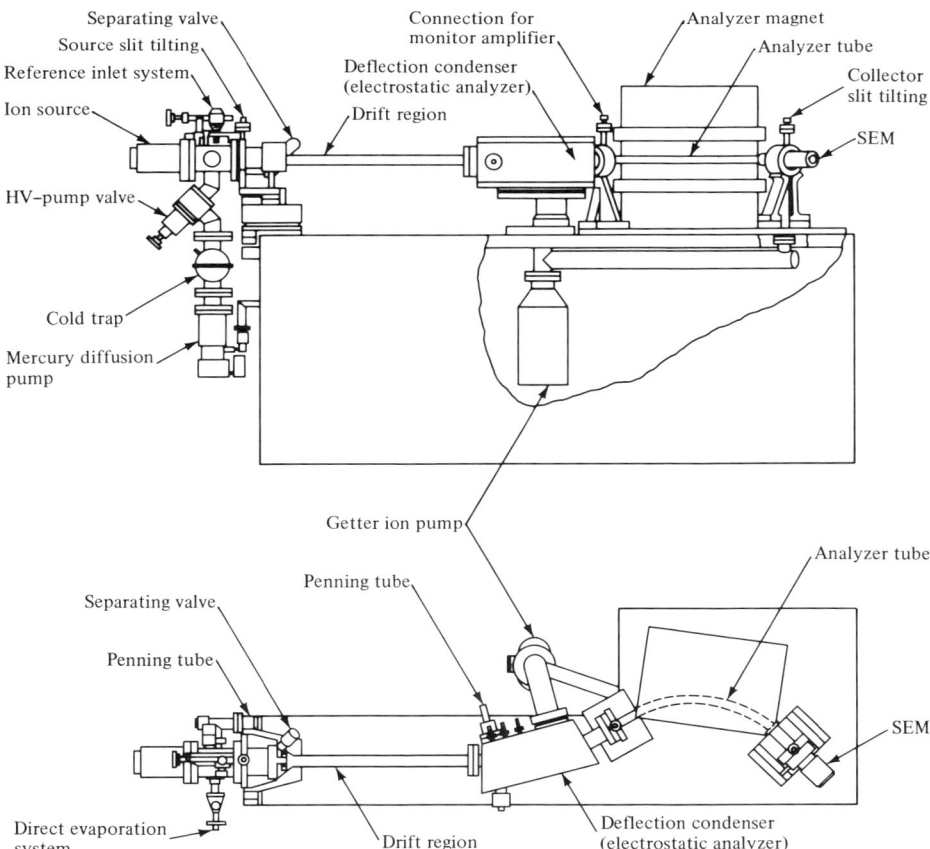

Figure 2-10 Schematic of a mass spectrometer (Varian MAT 711). (*From the instrument manual. Reproduced by permission of Varian MAT GmbH.*)

By placing a photographic emulsion, electron multiplier, or other recording device in the path of the dispersed beam, one is thus able to measure the relative abundance of ions of each m/e ratio. From this measurement is constructed a plot of abundance versus m/e, a *mass spectrum* of the molecule, and these plots (or the equivalent table) are the empirical data from which structure elucidation is attempted. Typical problems involve compounds with up to 50 or so nonhydrogen atoms. Larger molecules as a rule are not volatile and are thus not amenable to MS techniques in any event.

Figure 2-9 is a photograph of a mass spectrometer and Figure 2-10 is a schematic diagram of this particular instrument.

One of the convenient aspects of the technique is that less than a milligram (10^{-3} gram) of sample is enough; in some cases as little as a microgram (10^{-6} gram) is sufficient. The sample of the unknown compound is placed in the instrument by way of an inlet system and is vaporized because the sample must be in the gaseous state. A high vacuum is maintained in the instrument. Molecules diffuse through the opening into the *ionization chamber* where they are bombarded by electrons, causing some of them to ionize and break into fragments. The largest ion is usually just the total molecule less one electron (hence it remains positively charged but of the same mass as the molecule); this important fragment is the *molecular ion*, denoted M^+. Most molecules will yield a detectable signal for M^+. Ions that are positively charged will be accelerated (toward the right of Figure 2-10) by an electric field; the beam of ions is collimated by slits.

2.4.2 Structure Determination From MS Data

Masses of atoms are measured in *atomic mass units* (amu). An element may exist in different forms, called *isotopes*, that are chemically equivalent but of different masses. Usually one of these forms is far more abundant than the others. When speaking of a less abundant isotope, it is customary to signify its mass by a preceding superscript. Thus C normally denotes the most abundant isotope of carbon (or, generically, any isotope of carbon), while ^{13}C denotes the carbon isotope of mass 13 atomic mass units. Table 2-1 gives approximate atomic mass, the so-called nominal mass, for the most abundant isotopes of the elements mentioned in the examples that follow.[7]

Suppose our unknown compound produced the mass spectrum shown in Figure 2-11. We note first that the largest m/e is 18 amu. Typically, but not always, this largest value corresponds to the molecular ion, the fragment with the total mass of the molecule and a charge of 1. We can assume we have a compound rather than an element since there is more than one peak in the spectrum. We can now speculate as to what atoms comprise the molecule. We rule out any atoms with mass greater than 18. Possible candidates are then hydrogen, helium, lithium, beryllium, boron, carbon, nitrogen, and oxygen. We rule out boron, carbon, and nitrogen because we have no peaks at their masses: 11, 12, and 14, respectively. (The spectrum was not recorded below $m/e = 10$ in this example; in typical real problems the spectrum is not recorded below $m/e = 30$ or 40 because this portion is similar for most compounds and thus is not very informative.) The peak at $m/e = 16$ suggests the presence of oxygen. The peak at $m/e = 17$ suggests the fragment OH. Since the total mass is 18 and oxygen and hydrogen are

[7]Examples are from McLafferty, Benjamin/Cummings, Inc., 1967, Reading, Massachusetts.

Table 2-1 Nominal masses and valences of several elements

Element	Symbol	Nominal mass	Common valence
Hydrogen	H	1	1
Helium	He	4	0
Lithium	Li	7	1
Beryllium	Be	9	2
Boron	B	11	3
Carbon	C	12	4
Nitrogen	N	14	3
Oxygen	O	16	2

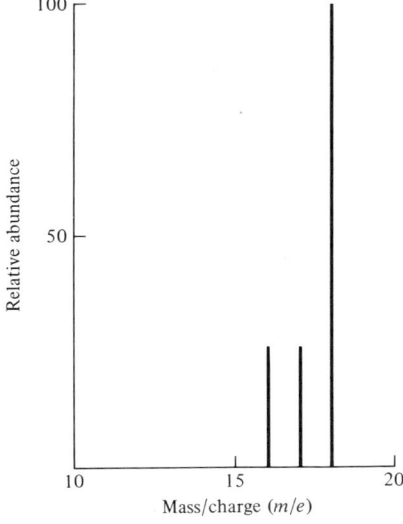

Figure 2-11 A mass spectrum from a simple molecule. (*Source:* Interpretation of Mass Spectra, *Copyright 1966, Benjamin/Cummings, Inc., Reading, Mass. Reproduced by permission. After McLafferty [1966].*)

present, helium, lithium, and beryllium cannot be present. We can conclude that the unknown was water.

A second example is given in the following tabular form. Following convention, relative abundances are normalized to percent of the highest peak.

m/e	Relative abundance
1	3.1
2	0.17
12	1.0
13	3.9
14	9.2
15	85.0
16	100.0
17	1.11
18	0.01

The peaks at $m/e = 1$ and 2 must be from hydrogen. The absence of peaks at 4, 7, 9, and 11 indicates no He, Li, Be, or B atoms are present. That being the case, the next peak at $m/e = 12$ indicates that carbon is one of the constituents. Peaks at 13, 14, 15, and 16 suggest that there are fragments containing one carbon atom and one or more hydrogen atoms: CH, CH_2, CH_3, and CH_4. The most likely hypothesis is then methane, with empirical formula CH_4. The peak at 17 may be explained as resulting from methane molecules containing ^{13}C. The small peak at 18 may be dismissed as due to an impurity (water).

A more complex example is the following.

m/e	Relative abundance
12	13.0
14	2.1
19	2.0
24	2.7
26	11.0
27	0.11
31	22.0
32	0.28
38	6.2
50	25.0
51	0.31
69	100.0
70	1.08
76	46.0
77	1.0
95	2.4
96	0.06

Suddenly our arithmetic puzzle has increased markedly in difficulty. We note that there are no series of several peaks differing by m/e steps of 1, unlike the previous example. This suggests that hydrogen is absent. The small peak at $m/e = 96$ can be assumed to be the molecular ion from molecules containing a heavier, less abundant isotope of some atom. If we assume that $m/e = 95$ is from the molecular ion, we can then assume that the peaks at 19 and 76 (=95 - 19) and at 26 and 69 (=95 - 26) are due to fragments of mass 19 and 26 and their residues. Now we need to do some guessing based on chemical knowledge. The mass of the most abundant isotope of fluorine is 19. The mass of 26 is quite apt to be from a fragment composed of one carbon and one nitrogen (12 and 14, respectively), a frequently occurring combination. The peak at 12 confirms the probable presence of carbon. There is a peak at 14 (nitrogen), although it is small. Candidate atoms are thus C, N, and F. Can these be combined in ways to account for all prominent peaks? Yes, for the only other prominent peaks are at 31 (=19 + 12) and 50 (=2 × 19 + 12). This suggests a structure CF_3-CN, a compound called trifluoroacetonitrile. Although this possibility may not be the only logical one, it is known to be a stable compound and is therefore a good guess, one that might find further support from other evidence. Note that the topological organization of these atoms suggests that fragmentations will yield pieces whose sizes correspond to the prominent peaks in the mass spectrum from which we began; for example, CF_3^+ and CN^+ are of m/e = 69 and 26, respectively.[8]

One can see that as the molecules become more complex, the problem of structure elucidation escalates drastically. Determining molecular structures from MS data is a complex art performed by experienced chemists and cannot be accomplished by the application of an algorithm to a set of data. We have outlined the process of analysis in these examples. From the spectrum one may find evidence for the presence of certain atoms or groups of atoms known to be stable. From other data and from knowledge of molecular structure and stability, one can rule out certain fragments and assign high probability to others. Hypotheses are then developed for the probable structure. One guesses how the hypothesized structure might fragment and then looks for the presence in the spectrum of m/e peaks corresponding to these ions. The search goes on until one or more structures are obtained that could have given rise to the spectrum. The problem is complicated by the fact that some peaks may be due to multiply charged ions; that some peaks have contributions from fragments containing low abundance isotopes of some elements; that the molecular ion may not be represented in the spectrum; that there are impurities in the sample that create noise in the data; that there are a great many possibilities to consider; and numerous other difficulties. There are other complications and refinements. We look next at some that are of importance for our purposes.

2.5 SOME IMPORTANT REFINEMENTS OF THE MS TECHNIQUE

2.5.1 High- and Low-Resolution MS

In the preceding examples we have spoken of atomic masses as though they were always an integral number of atomic mass units. This assumption is not so. By convention, the most abundant form of carbon is assigned a mass of exactly 12 amu. All other elements have nonintegral masses, as determined by careful measurements. Also, as we have noted, most elements occur in more than one form, and these different forms (isotopes) differ in mass. The mass of an atom, ion, or compound to the nearest integral value is called its *nominal mass*. The *accurate mass* of most elements is known to five or six significant digits. Table 2-2 gives values of accurate mass for the most abundant isotopes of some of the elements that are important in organic chemistry.

Rather than complicating matters, these nonintegral values are a great blessing, because an accurate molecular mass is a far less ambiguous indicator of composition than is nominal mass. See Lederberg (1964a). For example, CH_4 has an accurate mass of 16.0312 and oxygen an accurate mass of 15.9949; the difference is 0.0363. If a mass spectrometer can make sufficiently fine distinctions among (i.e., if it can resolve) m/e peaks and measure their positions precisely, it will be able to gather much stronger evidence about composition. For example, although there are four common possible contributors to a peak nominally at 28 amu, they differ slightly in accurate mass.

[8] A superscript "plus" denotes an ion of unit positive charge.

Table 2-2 Accurate masses of several elements

Isotope	Symbol	Accurate mass	Valence
Hydrogen	^{1}H	1.0078	1
Boron	^{10}B	10.0129	3
Carbon	^{12}C	12.0000	4
Nitrogen	^{14}N	14.0031	3
Oxygen	^{16}O	15.9949	2
Fluorine	^{19}F	18.9984	1
Sulfur	^{32}S	31.9721	4

Elemental composition	Accurate mass
CO	27.9949
N_2	28.0062
CH_2N	28.0187
C_2H_4	28.0312

Such slight differences can be measured easily if the contributors are resolved sufficiently. Low-resolution spectra provide information about nominal masses while *high-resolution spectra provide masses to an accuracy sufficient to assign elemental composition to each peak.*[9]

Referring again to Figures 2-9 and 2-10, note that ions of equal mass/charge ratio may enter the accelerating field with different velocities; they will thus have different velocities when they leave the accelerating field and enter the field-free drift region. Were this difference not compensated for, the magnetic field would not be able to bring into precise focus all the ions of the same m/e ratio and peaks would not be sharp and narrow. Instruments, known as double-focusing spectrometers, employ a "deflection condenser" to correct this aberration. The electrostatic field in this section exerts a *force* on an ion that is directly proportional to its charge but independent of its velocity. As a result, the ion is *deflected* from a straight path by an amount that is inversely proportional to its mass and the square of its velocity (i.e., to its energy) and directly proportional to its charge. The subsequent deflection in the opposite direction by the magnetic field exactly cancels the deflection in the deflection condenser and brings all fragments of equal m/e into focus. Thus the electrostatic section sorts ions

---

[9]By convention, two peaks are said to be resolved if the height of the valley between them is no more than 10 percent of their height. The *resolution of a mass spectrometer* is defined as nominal mass divided by resolvable mass difference. Thus an ion with a nominal mass of 360 and another ion of the same nominal mass that differs from the first only in that it contains an oxygen instead of a CH_4 will just be distinguished by an instrument with resolution 10000 (since 360/.036 = 10,000). Ions of larger mass that differ in the same way will be resolved less well (i.e., the valley between them will exceed 10 percent of their height). Low-resolution MS blurs these fine distinctions. The instrument depicted can record spectra of resolution 10000 up to m/e = 1500.

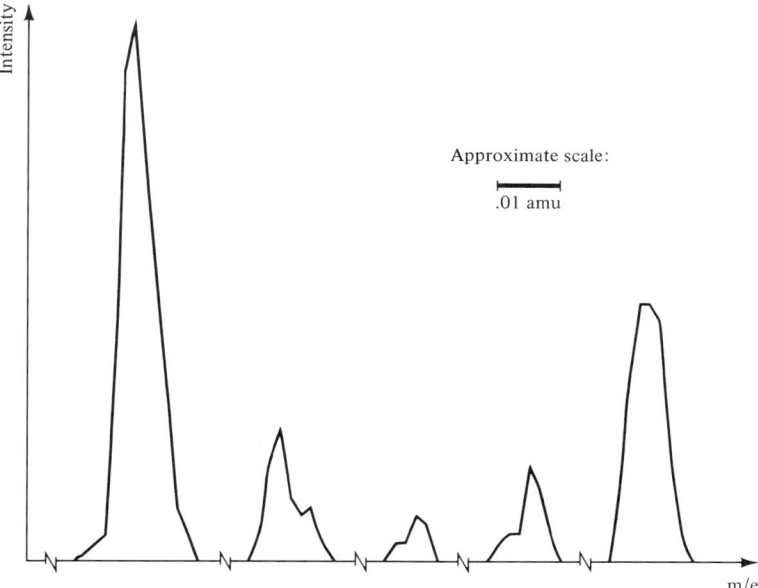

Figure 2-12 Sample of peaks from a high resolution mass spectrum before digitization: ion intensities plotted against m/e ratios of ions.

of equal charge on the basis of their energies; the magnetic section sorts ions of equal charge on the basis of their momenta. Ions of the same mass/charge ratio but different initial velocities will be path-separated when they emerge from the electrostatic section. A narrow slit at this position can then select only those of (nearly) equal velocity for entry into the magnetic section, which will deflect them by (nearly) equal amounts, resulting in a sharply defined peak accurately corresponding to their mass/charge ratio.

The double-focusing spectrometer is important when one needs to differentiate among various combinations of atoms possessing the same nominal mass. The bar plots (and tables) used to depict mass spectra in our examples are convenient means of recording low-resolution (nominal mass) spectra. A high-resolution (accurate mass) spectrum is recorded as a continuous curve consisting of peaks and valleys, produced from the output of the scanning electron multiplier. See Figure 2-12.

2.5.2 Metastable Ions

Ions spend very little time in the ionization chamber. In fact some fragmentation processes take longer than the fragment's tenure there. This fact means that further fragmentations of some ions take place elsewhere along the beam path of the instrument. Ions that fragment outside the ionization chamber are called *metastable ions*.

For reasons given in the following section, metastable ions prove to be important sources of data. If a parent ion of mass p fragments into a neutral fragment plus a daughter ion of mass d in the *drift region* after leaving the acceleration field and before entering the deflection condenser and magnetic section, the daughter will have been accelerated as an ion with mass p, but deflected as an ion with mass d. As a result it will not be recorded in the same position as it would have been had it come into existence in the ionization chamber. (If the fragmentation occurs in the deflection condenser or in the magnetic section, the daughter will not be recorded at all.) If the fragmentation occurs without loss of energy, the daughter ion will be recorded at the position where an ion of mass $m^* = d^2/p$ would be normally detected. [See Beynon (1972), pages 157–158.] Since the fragmentations are not without energy loss, and since the energy is released in random directions, the velocities of same-composition daughters are distributed about the zero-energy-loss values. This distribution results in a broad, relatively low peak centered at mass m^*, which is easily distinguished from "normal" peaks because of its breadth. To enhance the detectability of metastable peaks of low intensity, it may be necessary to widen the exit slit of the electrostatic section. This widening has relatively little effect on the intensity of sharp peaks, but relatively greater effect on the wide metastable peaks. This technique is called *metastable defocusing*.

When first noted, these unsharp metastable peaks were looked on as an anomaly of the method, and attempts were made to eliminate them. However, they are now recognized as the source of important information that cannot be obtained from normal spectral peaks. The instrument we have described is in fact designed to enhance the chances for the appearance of metastable ions and provide facilities for their measurement without interference from normal ions. The long field-free drift region between the accelerating field and the deflection condenser is to allow larger numbers of ions to decay during flight.

2.5.3 Rearrangement of Ions

The jumble of ions, hydrogen atoms, and neutral fragments in the ionization chamber provides opportunity for recombinations that produce ions that were not connected pieces of the parent molecule. Such *rearrangements* do in fact occur and complicate the picture further. The major value of metastable ions is that the presence of a metastable peak establishes a direct parent-daughter relationship between two ions since metastable ions are not formed in the ionization chamber and are thus not products of rearrangements. When this relationship exists, it is known that the two ions are connected (or overlap) in the parent molecule.

A very common rearrangement is called *hydrogen transfer*. This arrangement occurs when hydrogens migrate from one ion to another resulting in a more stable structure. As it turns out, such transfers frequently cause the appearance of peaks 1 or 2 amu less than the expected peak. The patterns of transfer are to an extent consistent, and knowledge of these patterns is important in the interpretation of mass spectra.

Another important example is the *McLafferty rearrangement*, which is any of a class of rearrangements in which a hydrogen atom transfers to a polar,[10] odd-electron ion during the fragmentation process. The peaks produced by such rearrangements fre-

---

[10] A polar ion is one whose charge is localized so that the electric field of the ion is not symmetrical.

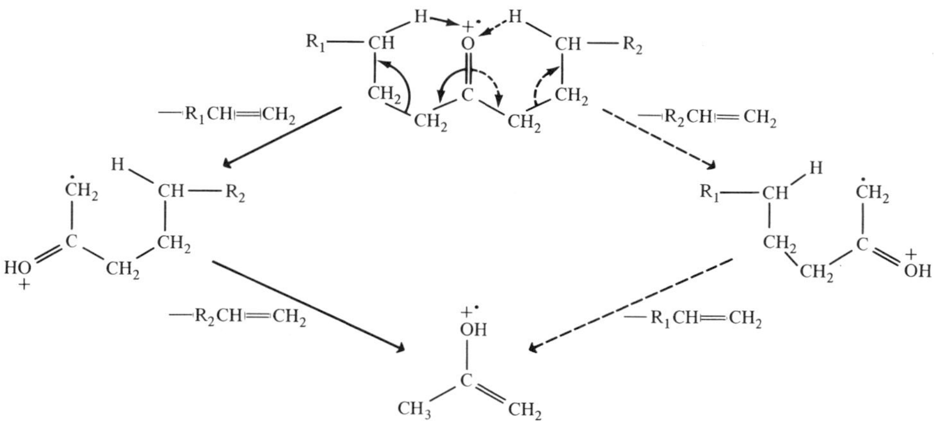

Figure 2-13 McLafferty rearrangements. (*Source:* Mass Spectrometry: Techniques and Applications, *edited by G. W. A. Milne, Copyright 1971, John Wiley & Sons, Inc. Reproduced by permission. After Buchanan, Duffield, and Robertson [1971].)*

quently are quite prominent. Figure 2-13 depicts a complex rearrangement process involving four occurrences of McLafferty rearrangement. Metastable peaks corresponding to each of the four transitions might well appear, as well as peaks corresponding to the distinct ions.

2.5.4 The Nitrogen Rule

Of the elements that predominate in organic compounds, nitrogen is unique in having an even nominal mass (14) and an odd valence (usually 3, occasionally 5). A moment's reflection will demonstrate that any hydrocarbon will have an even nominal mass, unless it contains ^{13}C. Thus any *ion* containing only hydrogen and carbon that results from breaking one single bond will be of odd nominal mass. The picture is not changed if we include sulfur, with even nominal mass (32) and valence (4). Thus if there are no nitrogen atoms in the compound (and no phosphorus—mass 31, valence 3), all large ions will be of odd mass, except those that involve the loss of an odd number of hydrogens or breaking of double bonds.

If a single nitrogen atom is contained in the compound, the breaking of any one of its bonds will yield an odd mass ion and an even mass ion (containing the N). Typically, then, if there are no nitrogens, the major peaks will all be of odd mass, while if there is one nitrogen, there will be major peaks of even mass.

The reasoning generalizes. An ion containing an odd number of nitrogens (and no double bonds) has an even mass; an ion containing any even number of nitrogens (and no double bonds) has an odd mass. It follows that a molecule with no nitrogens will produce only peaks of odd nominal mass if we ignore the breaking of double bonds and multiple breaks. A molecule with exactly one nitrogen atom will produce both odd and even nominal mass peaks. If a molecule contains more than one N (whether an even or odd number), it can produce both even and odd mass fragments. However, if the nitrogen parity is even, it will tend to produce more odd peaks, and conversely. In particular, if there is a series of even peaks in a low-resolution spectrum that differ by multiples of 14 amu, these differences probably correspond to ions that contain an odd number of nitrogens and differ in their count of CH_2 radicals, a frequently lost structure. If there is a series of odd peaks that differ by multiples of 14 amu, these differences probably correspond to ions that contain an even number of nitrogens (or none) and differ by their CH_2 count. Thus the nitrogen parity of a molecule can be inferred by the parity of the CH_2 series: odd series, even nitrogen parity, and conversely. This rule is the *nitrogen rule*. It is not infallible, because double bonds and hydrogen transfers can cause the appearance of other peaks of substantial relative abundance. The rule is not important when high-resolution spectra are available, since exact compositions are then known in any case.

2.6 OTHER ANALYTICAL METHODS

A chemist can gather other kinds of information about an unknown compound. At the very least there will be some information about its source and extraction methods, which often suggests probable constituents. Laboratory analyses may also provide further clues of this sort. In addition other instrument-based techniques provide information about the compound. The most common among these are *gas chromatography*, *infrared spectrometry*, *ultraviolet spectrometry*, and *nuclear magnetic resonance spectrometry* (NMR). At the time of writing DENDRAL has not been extended to the interpretation of these data [with one exception, Carhart and Djerassi (1973)] in the way it interprets MS data. However, this information is employed indirectly in the procedures that guide the DENDRAL search through the set of possible structures. The following brief descriptions give some idea of the nature of the information provided by these other techniques.

2.6.1 Gas Chromatography

This procedure is very important in analytical chemistry. The term *chromatography* is historical and can be misleading since most forms of chromatography do not involve color. A gas chromatograph is a device that separates a mixture of compounds into its constituents by taking advantage of the differences in partitioning of different compounds between a moving gas stream (the mixture to be analyzed) and a stationary phase (a viscous liquid). Different constituents interact differently with the stationary phase; in particular more volatile components move through more rapidly and thus exit the chromatograph earlier. By collecting and measuring the effluents one can determine their approximate relative abundances, and relative boiling points. Gas chromatography is frequently a useful precursor to MS as a means of extracting pure samples from a mixture of compounds. Such an arrangement is referred to as *GC/MS*.[11]

[11] Gas chromatography is the source of many of the high- and low-resolution mass spectra analyzed by chemists working with DENDRAL. A separate data acquisition system has been developed for a Digital Equipment Corporation PDP-11 computer to collect GC/MS data. It includes a program, named CLEANUP, that provides mass spectra of the individual components of mixtures, such as urine [Dromey et al. (1976)].

2.6.2 Infrared Spectrometry

Just as a piece of colored glass differentially absorbs visible light of different wavelengths, a sample of organic compound differentially absorbs electromagnetic radiation of different wavelengths. From these patterns of absorption some information about the compound can be obtained. Wavelengths between 2.5 and 15 microns are particularly useful. This radiation is in the infrared region of the electromagnetic spectrum.

The reason that radiation is absorbed by matter is that the energy of the radiation alters the matter in some way. Since the alterations must be by discrete (quantum) amounts, the energies absorbed are of particular wavelengths. Wavelengths in the infrared region alter molecules by increasing their rates of vibration and bending. Vibration refers to oscillations in the effective lengths of bonds (as though the atoms were balls connected by springs), and bending refers to oscillations in the bond angles. Since different types of bonds have different characteristic rates for these oscillations, the absorption spectra for infrared radiation yield information about the types of atoms, and hence the types of compounds, that are present.

This technique can not be used alone to determine the structure of complex organic compounds, but it is useful for detecting the presence of certain functional groups. Thus it is useful in conjunction with mass spectrometry, which, as we have seen, is benefited by any hypotheses about what components are apt to appear as fragments in the mass spectrum.[12]

2.6.3 Ultraviolet Spectrometry

Another band of the electromagnetic spectrum that is differentially absorbed by organic compounds is the range of wavelengths from 0.01 to 0.38 micron (10 to 380 millimicrons), in the ultraviolet region.

This radiation is of much higher energy than infrared, and the alterations it effects on the molecules irradiated are of an entirely different nature. Ultraviolet radiation causes quantum shifts in the energy states of orbital electrons. Since energy states differ for different types of chemical bonds, the energy needed to cause such shifts is different for different compounds, being dependent mainly on the arrangement of double and triple bonds. By observing which wavelengths are absorbed, the chemist can infer, in some cases, some of the functional groups of the complex molecules of the sample. In general, the information obtained in this way is complementary to that obtained from infrared spectrometry.

2.6.4 Nuclear Magnetic Resonance Spectrometry

Nuclear magnetic resonance spectrometry can be viewed as another technique involving differential absorption of energy. In this case the energy is supplied by a magnetic field that oscillates at radio frequency.

To understand this technique it is necessary to look inside the atomic nucleus. The protons in the nucleus are spinning, and since they are charged particles, their spinning causes them to become tiny magnets (magnetic dipoles). If a large steady

---

[12] Computer programs written by Woodruff and Munk (1977) analyze infrared spectra in much the same way that the DENDRAL planning program analyzes mass spectra (see Chapter 5).

magnetic field H_0 is applied to a sample of a compound, the spinning protons in the nuclei have only two stable orientations with respect to the direction of H_0: either their magnetic fields are parallel to H_0, or they are antiparallel[13] to it. The former state is the lower-energy and hence more stable state, and the majority of protons will be so aligned.

Actually, the protons in the lower-energy state will not be exactly aligned with H_0. Rather their axes of rotation will precess about the direction of H_0. Imagine a spinning top that has lost energy and is slowing down. Its axis will precess about the direction of the gravitational field with a certain angular velocity (that changes as the top slows). If we were to tap the side of the top perpendicular to its rotational axis at just the right frequency (that is, if our tapping were in *resonance* with the frequency of precession), the precession angle would increase until the top fell over. This imperfect analogy describes what happens to protons in a nuclear magnetic resonance spectrometer. An oscillating magnetic field H_1 at right angles to the fixed field H_0 acts as the tapping finger. Instead of falling over, the protons flip to their higher-energy state. Since the angular velocity of precession depends on the magnitude of H_0, the correct tapping frequency also depends on H_0. It is because the protons absorb energy of a particular magnitude when flipping to their higher-energy state that this technique is a form of spectrometry.

But what makes this energy absorption a function of chemical structure? Since the nuclei of atoms are surrounded by orbiting electrons, and since these moving charges set up magnetic fields counter to that of the applied field H_0, the electrons shield the protons from the effects of H_0 and H_1. The effectiveness of the shielding depends on the details of the motion of the electrons. But these motions in turn depend on the atoms present (in particular whether an atom has an even or an odd number of electrons), and on the type of bonding, since bonding is effected by sharing of orbital electrons in complex ways.

The result is that shifts in absorption frequency with respect to a fixed standard compound enable the detection of certain substructures in an unknown compound, which, again, is the information needed to help interpret mass spectra. The first NMR techniques to be developed detected hydrogen nuclei (protons). In particular, NMR can be used to determine the number of carbon-bonded protons (hence the number of methyl radicals) in a compound, and also the number of nitrogen-bonded protons. It is now possible to gather information about the nuclei ^{2}H, ^{11}B, ^{13}C, ^{15}N, ^{29}Si, and ^{31}P.

2.7 LIBRARY SEARCH

With each of the spectroscopic methods just discussed, a standard method of interpretation is to compare the spectrum of the unknown compound against a library of spectra of known compounds [McLafferty and Venkataraghaven. (1978), Heller et al. (1977)]. Because of experimental variables, an exact match is seldom expected. However, numerous closeness-of-fit criteria have been developed to find the "closest" match to the unknown in the library.

---

[13] Antiparallel means parallel but of opposite polarity.

A library of mass spectra is routinely checked as part of the data acquisition system used to collect spectra for Stanford University chemists and for DENDRAL. The spectra found in the library do not require further interpretation.

2.8 SUMMARY

We have described the structure elucidation problem, which is ubiquitous in modern chemistry. In solving such problems, the chemist may have access to useful information from many sources, including data from the several powerful analytic methods just described. Matching new data against libraries is a standard method of structure identification. However, when new compounds are encountered, as frequently happens,[14] library matching is insufficient and the chemist must interpret the data from sources such as the ones described here. These various sources of information provide different types of facts about the structure of the molecule being investigated. All, however, can be cast in terms of constraints on the connectivity of the molecule's constituent atoms. The chemist thus is faced with a sort of jigsaw puzzle and has to fit together the various facts into a coherent "ball-and-stick" picture of the molecule.

---

[14] New chemical compounds are discovered at the rate of about 100,000 per year.

CHAPTER
THREE

ARTIFICIAL INTELLIGENCE

Artificial intelligence is that part of computer science that studies computational methods for complex symbolic (not necessarily numerical) problem solving. Such mechanization of symbolic reasoning stands in marked contrast to traditional formal methods of problem solving used in science and mathematics. The major approach of artificial intelligence is heuristic programming, which replaces exhaustive enumeration of cases with selective consideration of alternatives. DENDRAL applies a specific heuristic programming paradigm to the structure elucidation problem and to the task of hypothesis formation.

3.1 INTRODUCTION

The DENDRAL Project spans approximately half the history of artificial intelligence (AI) research. This period has been one of great change in computer technology and in the attitudes, aspirations, and activities of the research community working on problems of artificial intelligence. To place DENDRAL in the context of its parent discipline so that we may see how it has drawn from and contributed to this development, we first look at the short history of AI.

The major goal of AI research is a productive understanding of the processes of intelligent thought. The major method of AI distinguishes this discipline from others with similar goals. That method is the creation of intelligent artifacts, currently in the form of computer programs. As yet the field has little of what could be called formal theory, and thus to some it appears to be a curious gallimaufry of hardware technology, software packages, programming tricks, specialized problem-solving procedures, information representation schemes, human intuition, examples of things that do not work, game-playing strategies, mathematical theorems, debugging techniques, primitive robots, and endless specialized knowledge at many levels of abstraction. A closer look reveals, we believe, an emerging discipline with a measure of cohesiveness of concepts and techniques.

In its first decade (approximately 1954-1964) AI was characterized by enthusiasm and optimistic forecasts of the imminent solution of many problems of psychology, linguistics, mathematics, philosophy, technology, and management. By the end of the decade, when DENDRAL was in its initial stages, these optimistic forecasts were being reevaluated. The now classic example of unfulfilled forecast is mechanical translation of natural languages. This and most other problems outside previously formalized domains have proved far more difficult than many AI pioneers imagined. From the vantage point of perfect hindsight it seems remarkably bold to have forecast immediate solutions to problems with which other disciplines had struggled for decades or centuries. But from the less humble perspective of a young idea, the possibilities did indeed look promising. Many of the abilities to be automated were, after all, commonplace things done by ordinary people. Why, every five-year-old can speak at least one language! Technology had brought us from horseback to supersonic flight, from pony express to satellite video links, and from steam engine to atomic energy in less than a century. With that acceleration and the newfound power of computation, solutions to the problems of human thought, though they had a long history, might well have been just around the corner.

In the mid-1960s AI was in a transition stage. On the one hand the newly discovered difficulty of the adopted problems was tempering optimism and in some quarters producing pessimists. Long-standing critics of determinism in general, and technology in particular, were coming out of the closet. On the other hand, some work on general problem-solving procedures still held promise in the view of many researchers. The General Problem Solver (GPS) [Ernst and Newell (1969)] had yet to have its limitations established. A bright light was the resolution procedure [Robinson (1965)] for proving theorems in the predicate calculus: a complete, uniform proof procedure for a general calculus seemed promising indeed. Ironically, the search for general problem-solving methods was being pursued on a broad front at the same time that specific solutions to special problems were failing to produce results.

A second aspect of this transition stage was the astoundingly rapid development of computer technology. Solid-state, second-generation hardware was replacing the slower, less reliable equipment. Telecommunications technology was being interfaced with computers on a large scale. Time-sharing of computers had been conceived and was to become a reality in the 1960s. Software in general and executive systems in particular were reaching new levels of sophistication. List-processing and string-processing languages were becoming readily available. These developments combined, in the early 1970s, into facilities that eclipsed in power those available 10 years earlier. Today's student using a high-level language on an interactive terminal is interminably amused by tales of the days when a programmer would carry a box of punched cards to a computing center and return some days later only to find that a bug in the compiler, a mispunched control card, or an operator error had terminated the job. Although none of today's wealth of computer sophistication would have struck the early workers as science fiction beyond their dreams, the additional load imposed by the bothersome conditions of the time were enough to swamp many well-conceived projects of realistic scope.

It was in this milieu that DENDRAL began. An important fact about DENDRAL, though not unique to it, is that it undertook a relatively narrow and well-defined problem for which there was a clear measure of success. To this problem it applied specific, task-directed methods and knowledge. The major lesson DENDRAL has for artificial intelligence, and for those disciplines interested in the application of AI techniques, is that it is possible to select problems of modest complexity that nonetheless baffle the novice, and to reduce these problems to some order, resulting in a problem-solving system that lends needed assistance to human intelligence. By lowering one's sights from solving broad, general problems to solving a particular problem, by applying as much specific knowledge to that problem as can be garnered from human experts, and by systematizing and automating the application of this knowledge, a useful system can be produced.

3.2 PROBLEM-SOLVING METHODS

We use *problem solving* as a generic term for a broad class of cognitive activities investigated by AI. Other terms would do as well and many have been used in that capacity from time to time. The field is not yet plotted sufficiently well to draw sharp distinctions between question answering, information retrieval, pattern recognition, intelligence engineering, concept formation, learning, induction, abstraction, hypothesis formation, or any of a number of other processes that have been suggested simultaneously as manageable subproblems and exhaustive paradigms. We rather arbitrarily prefer our unassuming generic term. What we embrace under the problem-solving label is a specific set of programs, real and imagined, that have been produced or proposed by AI researchers.

Digital computers are most naturally suited for manipulating finite data structures composed of tokens from a finite alphabet of discrete symbols. The simplest such data structures are numbers. Formulas of a formal language are a more general and relatively common example. Arbitrarily complex graph structures are frequently manipulated by AI programs. For a machine that deals only with such objects, we must define *problem* and *problem solving* in terms of finite structures and transformations on them. In general, problem solving may be thought of as taking place in a *problem space* whose points, called *problem states*, are defined in a language for representing data structures [Nilsson (1971)]. Goals are distinguished problem states, and solving a problem means locating a goal state. Accordingly, we introduce the following *problem-solving paradigm* in order that we may properly characterize the DENDRAL problem-solving methods.

(1) A *problem-state language* is a formal language that defines a class of data structures constructed from discrete symbols. Each expression in such a language defines a potential *problem state*. (2) There are *transformations* that are rules for changing one problem state into another. Together, the set of problem states and transformations determine a *problem space* whose connectivity is defined by the transformations.

(3) A *goal state* is a distinguished problem state. (4) An *initial state* is a distinguished problem state; any problem state may serve as an initial state if it is assumed that it can be reached (constructed) a priori. (5) The space may or may not have other properties. For example, it may have a metric defined on it; it may be dense, it may be continuous, and so forth. (6) There may be one or more *secondary problem-state languages* in which descriptions of the states can be expressed. These descriptions are in terms other than those that directly define the connectivity of the space, sometimes in a well-constructed language of abstractions. (For example, in chess, piece advantage is a description that is not a defining characteristic of board configurations or legal moves.) Transformations may or may not be defined for these languages. (7) A *problem solver* is a procedure that attempts to find one or more goal states.

Though it cannot be shown that this conception of problem solving encompasses all problem-solving activities, it is in fact very general. Most board games readily fall into this scheme: initial states are board configurations, the legal moves define changes in board configurations, and goal states are positions that meet certain formally defined critieria. A player qua problem solver seeks a sequence of moves that takes him from initial position to a win. Theorem proving is problem solving that seeks a path of legal inferential steps from axioms to theorem. Symbolic integration and similar formal mathematics is a matter of finding a sequence of legal transformations that lead one from integrand to integral, from equation in unknown x to an expression defining the value of x.

Finding the maximum of a function is a search through the continuous metric space defined by the variables, looking for a point at which the function has its greatest value. Repair of a television set requires finding a set of realizable modifications that will convert the set that malfunctions in a particular way (the initial state) into a functioning set (the goal state). Finding a misplaced object involves thinking of all places it might be and looking at each one (no initial state); the goal state is the place where the object is. Building a house is a problem of finding a sequence of possible building activities, defined by the functions of the available tools, that leads from a pile of materials to a house-shaped arrangement of materials. Science as problem solving is the search through the space of possible explanations to find one that meets criteria of parsimony, empirical validity, and usefulness; the grammar of a formal or natural language defines the space of possible explanations.

However, simply because a problem can be seen to fit the general paradigm as we have defined it does not mean that seeing it in this light is helpful. Carpenters would find our analysis of their problem of little use. Scientists would find the above formulation of their problem a poor characterization of their own activity.

Furthermore, an intuitively conceived problem can always be forced into the paradigm in a number of ways and the representation chosen will often make the difference between trivial and impossible. We know of no nontrivial exceptions to the claim that all AI problem-solving approaches have finessed this *representation problem* by leaving it to the human programmer/scientist.

Nonetheless there are many problems that fit naturally into the paradigm, and for which an appropriate representation is available. For these it is yet necessary to define the methods of exploring the space. A number of them have been studied. The analysis of these methods and the related problems comprises an important part of AI theory. It will be clear that different problem-solving strategies are appropriate for different problems.

3.2.1 Nonheuristic Methods

Random exploration This method is random selection of problem states (with replacement); the probability of selection is uniformly distributed over the set of accessible states. For large spaces the random procedure is a last resort. It may however be the best strategy for some problems, such as finding a cross-eyed tiger in the jungle or looking for an adaptive mutation. For difficult problems this method offers little hope, and, in general, effective problem solving requires the use of additional knowledge. A variation on random search is to employ a nonuniform probability distribution that makes use of some knowledge about the distribution of solutions.

Systematic exhaustive exploration Searching the space exhaustively and nonredundantly (i.e., without replacement), is superior to random search if the memory and time costs of the additional computation for record keeping are not excessive. It is the method of choice for finding needles in a haystack. However, in some cases this procedure may work less well than random search if the system of searching tends to confine activity to a localized region of a large space.

Algorithmic methods An algorithm for a class of problems is a procedure that is guaranteed to find a solution to any problem in the class if a solution exists (this property is called *completeness*), that will indicate that no solution exists when that is the case, and in either case will terminate in finite time. For some classes of problems it is possible to construct an algorithm based on systematic exhaustive exploration. However, "good" algorithms generally do not naturally fall into the paradigm of problem-space exploration because there is a directness and efficiency about their discovery methods. For example, in differentiating a polynomial there is no explicit consideration of the space of all polynomials; the correct one is "computed directly" from the given expression. On the other hand, collecting terms and simplifying the answer after differentiation clearly has the character of choosing among possible transformations and considering alternatives in search of the one that meets certain criteria of, for example, simplicity.

3.2.2 Heuristic Methods

Cousins to algorithmic methods are heuristic methods. A heuristic program for a class of problems is usually defined as one that does not guarantee a solution to every problem in the class, or that has no known bounds to its inefficiency. We hope these liabilities are offset by a measure of efficiency in solving interesting and important members of the problem class.

Heuristic programs characteristically have the flavor of exploration that is suggested by our problem-solving paradigm. In addition, however, the exploration is selective rather than exhaustive.

We distinguish two types of problem-space exploration that we call *search* (not to be confused with exploration in general) and *generation*. Search is the production and examination of problem states. Generation is the production and examination of candidate solutions.

3.2.2.1 Heuristic search Those search procedures whose performance on a class of problems cannot be guaranteed, either as to completeness or as to cost, but that make use of knowledge of the problem space over and above its abstract definition, are called heuristic search procedures. They are of many varieties, characterized by the kind of additional knowledge of the problem space on which they attempt to capitalize.

Statistical Our paradigm permits that problem states may sometimes be characterized by certain properties or features, stated in what we have called secondary problem-state languages, that add a new set of dimensions to the space independent of the defining transformations and formulas. When such problem-state descriptors exist, particular features of the initial states might be associated for some reason with particular features of the goal states for problems of interest. Statistically guided search attempts to capitalize on this situation, either by employing known associations or searching for some.

Hill climbing Hill-climbing methods (1) select a point in the space, (2) search in the local area of that point for the direction that maximizes a gradient, and (3) move in the direction of maximum gradient to a new point where the process is repeated. The name derives from situations in which the problem is to find the maximum of a continuous function of one or more real variables. In the case of a continuous function of two variables, the plot of the function is a surface and the problem is to find the peak of the highest hill on that surface. By moving in the direction of steepest ascent (readily determined if the function is differentiable), the search climbs to the top of the nearest hill. The strengths and weaknesses of this method are readily apparent from this image. Local maxima (peaks and mesas) capture the climber who can not descend to find a higher hill. Methods to select starting points are crucial, therefore, and it is advisable to use more than one starting point. Also critical is the step size: too small is inefficient and too large may step the climber past the peak, so it may be wise to vary step size as a function of the current position and the selected gradient. Likewise one cannot afford to compute gradients in all directions, so selection is necessary here too.

Abstraction planning It may be possible to take a coarser view of the problem space, as though we could back off and see the general features while omitting the details. This is one form of planning (a different form than embodied in Heuristic DENDRAL, to be described shortly). We will call it *abstraction* to keep the terminology clear.

In the terms of our paradigm of problem solving, abstraction can be achieved by defining a new set of problem-state descriptions, based on a language that is an abstraction of the basic language in the sense that one problem-state description in the abstraction language corresponds to many problem-state descriptions in the basic language. The new view of the problem space now offers a diminished set of states and therefore reduces the complexity of the problem. Transformations for the abstraction language define the connectivity of the abstracted space. If the problem solver can discover a path from initial state to goal state in the abstracted space, he has not solved the original problem but has established a plan. Each step of the plan then becomes a problem in the original space, but the combined complexity of all these problems may be less than the complexity of the original problem.

Working backward For some problems the number of alternatives to be searched is fewer if we begin at the goal state and, running the transformations in reverse as it were, search for the initial state. This situation might be the case if there is only one goal state but a multitude of initial states. Working backward is a time-honored procedure in mathematics and logic. Of course it is of no use for problems in which the transformations are not reversible, or when the definition of goal state is in terms to which the transformations cannot apply, as with chess. One would be hard put to play chess by working backward from a definition of checkmate, looking for the initial board configuration.

3.2.2.2 Heuristic generation   *Generate and test*
If there is a procedure that can generate candidate *solutions*, goal states in our present terminology, it may be possible to solve problems by the sequential enumeration and checking of potential solutions. This method is frequently called generate and test. Note that in this paradigm it is not the *states* of the problem space that are generated. Indeed there need not be a problem space of the sort we have been discussing. Here we need only a space of solutions that can be generated for consideration. This paradigm is thus fundamentally different from searching the space of problem states (state-space search).

Analogs of the heuristic search methods described above apply to heuristic generation. Statistical information concerning the distribution of solutions (perhaps by categories defined in a "secondary solution description language") may be used to guide generation. If some measure of goodness is computable from a proposed solution, then a hill-climbing procedure could be used to determine ways of modifying one proposal in appropriate ways. Similarly, if an abstracted description of the set of solutions can be constructed, it may be possible to search first for the correct solution class, and then search within that class.

These methods do not exhaust the possibilities, but they provide enough structure for our discussion.

3.2.3 Multiple Sources of Knowledge

Much more problem-solving power can be achieved if there is more than one source of information that can be used. For this reason purely syntactic problem solvers are inherently less powerful than those that employ semantic information as well. The secret is not that semantic information is more important, but that two sources of guidance are better than one.

Jigsaw puzzles are an appropriate image to make this procedure clear. The problem of finding the one arrangement of all the pieces that yields the desired picture may be a very difficult and large combinatorial task. If the puzzle were done with the pieces face down, analogous to having only the "syntactic" information of piece contours, the true difficulty of the problem would become apparent. If all the pieces were of the same shape, squares or hexagons, the problem could only be solved from the "semantic" information of colors and pictured objects and would also be very difficult. Jigsaw puzzles are tractable because both these sources of information are available and can be played off against one another.

In terms of the paradigm of state-space search, different sources of information correspond to different problem-state languages. Several of the search heuristics defined above rely on these secondary languages. Hill climbing needs a language in which to define its gradient of "warmth," and statistically guided search is based on a rather general but blurry-visioned search for descriptors that have some correlational information. Abstraction planning is another way of bringing to bear different views of the problem space.

On a more general level, apart from any of these methods, is the representation problem. A problem has in general many possible representations. It may be possible to choose two or more (rather than just *one*) in such a way that progress in one state-space search can be transferred to another. Thus those transitions that are difficult in one representation may be bypassed by using a second and vice versa.

In the context of heuristic generation, multiple sources of knowledge have the effect of limiting generation to the *intersection* of the solution sets delimited by each source.

3.3 DENDRAL

DENDRAL is not a single program but a set of programs. Some of these programs may be used alone to perform single subtasks of importance to the problem of chemical structure elucidation. Some may be linked in various ways by different executive programs to form coherent systems for doing larger tasks. To organize the description of this collection of intertwined programs we first note that they comprise basically two systems. The first, called Heuristic DENDRAL, is a system that incorporates specific knowledge of chemistry and mass spectrometry, accepts a mass spectrum and other experimental data from an unknown compound as input, and produces an ordered set of chemical structure descriptions hypothesized to explain the data. The second system, called Meta-DENDRAL, accepts known mass spectrum/structure pairs as input and attempts to infer the specific knowledge of mass spectrometry that can be used by Heuristic DENDRAL to explain new spectra. Heuristic DENDRAL is a performance system and Meta-DENDRAL is a learning system.

The following chapters describe these two systems. It must be kept in mind that each has evolved over many years; in fact each is being continually revised. The descriptions here are somewhat idealized expositions of the programs as they exist at this writing. Earlier versions, described in other publications, were used to produce many of the results summarized later. No single version has been systematically applied to all the specific structure elucidation problems investigated by the DENDRAL project.

Naturally, neither Heuristic DENDRAL nor Meta-DENDRAL was developed by following a detailed blueprint established in advance. The design process inevitably is one of trial and error and revision. The descriptions that follow organize these programs around conceptual paradigms that themselves are products of the development process. Early systems, and earlier publications, were conceived less clearly and described in somewhat different terminology, although the current vision, in retrospect, encompasses these earlier designs.

Two important features of the DENDRAL system characterize it and set it apart from most other AI systems. The first is the basic organization of the problem-solving method, which we have called the plan-generate-test paradigm. The second is the fact that it incorporates and gains its heuristic power from considerable task-specific knowledge overlaid on a general syntactic method; we call DENDRAL a knowledge-based system for this reason.

3.3.1 The Plan-Generate-Test Organization of DENDRAL

The basic method of both Heuristic DENDRAL and Meta-DENDRAL is an important extension of the generate-and-test paradigm. The heart of this paradigm is a *generator*. This is a program that enumerates for a particular problem its potential solutions, which are expressed as chemical graphs in the case of DENDRAL. It is often desirable, though not essential to the paradigm, for the generator to be exhaustive and nonredundant, that is, that it guarantee that every possible solution will be enumerated exactly once. The Heuristic DENDRAL generator has these properties. The Meta-DENDRAL generator does not, since the set of all possible solutions is not well defined.

When there are a large number of candidate solutions, success will be rare unless we can limit generation to the most likely candidates. DENDRAL's way of limiting generation is with a planning program that can suggest constraints on generation. This component distinguishes plan-generate-test from generate and test. Constraints may take the form of ruling out large sets of candidate solutions or suggesting exhaustive search over limited classes of solutions, or both.

The DENDRAL *planner* is a hypothesis-formation program that employs task-specific knowledge to find constraints for the generator. It is important that the planner be extremely flexible in the sense of permitting the ready addition of new knowledge. Ideally, the knowledge will be highly modular so that it is possible to add new knowledge without reevaluating the old. A key feature of the plan-generate-test paradigm is the interface between planner and generator: the output of the planner must be in a form appropriate to the language of the generator.

In DENDRAL, having narrowed the search space by planning, the generator proceeds to produce all and only those solutions consistent with the plan. Typically this set of solutions, though much smaller than the entire space, will be undesirably large. Further selection is performed by the third and final stage, the *tester*. This is a program that examines each proposed solution and rejects those that fail to meet certain criteria. The tester incorporates a theory of mass spectrometry that predicts what fragmentations a proposed chemical structure will undergo in a mass spectrometer and constructs a mass spectrum accordingly. This predicted spectrum may then be compared to the one produced in the laboratory. The DENDRAL tester programs are usually referred to as PREDICTOR programs because of their special form.

Both a planner and a tester are programs for constraining the set of likely solutions. It might be thought that it is much more economical to apply constraints first, in the planning stage, rather than last, in the testing stage. In some measure this is true, and every effort is made to preconstrain the generator. However, preconstraint is not always possible nor desirable, as we shall see in detail later.

The feature that gives the plan-generate-test paradigm its cohesiveness is the uniform representation used by the three components. In the case of DENDRAL this representation is chemical graphs. The planner devises hypotheses that reject and/or propose certain classes of chemical graphs, the generator generates chemical graphs, and the tester represents fragmentation processes in terms of chemical graphs. This common representation is the glue holding DENDRAL together.

Uniform representation plus the plan-generate-test paradigm are the keys that permit multiple sources of knowledge to be brought to bear within the generate-and-test method. The ways in which generation is constrained may be of fundamentally different types. In fact in the case of DENDRAL, the knowledge from which constraints derive may be from any source as long as it can be translated into appropriate constraints on graph generation. DENDRAL's power derives in large part from this ability.

DENDRAL is a particular instantiation, or rather several particular instantiations, of the plan-generate-test paradigm. Numerous variations are possible. The paradigm is still applicable and potentially powerful even in the absence of a nonredundant or a nonexhaustive generator. If the space of solutions is not large, either the planner or tester may be omitted. Without either the planner or tester the generator alone may still provide useful information by providing a measure of the size of the set of possible solutions to a problem, or a relative measure of the difficulty of two problems. Further, the generator may be used in conjunction with a human problem solver if the human is able to define the constraints. This is the case with CONGEN, the CONstrained GENerator program, described in Chapter 4. Many variations on this theme are possible.

3.3.2 Knowledge Engineering

We have noted that specific knowledge is used by DENDRAL to constrain the generation of solutions and to test proposed solutions. The plan-generate-test paradigm does not require this approach, and certainly other paradigms reject it. As we noted earlier in this chapter, many of the early AI projects, as well as other approaches to a theory of intelligence, have sought general problem-solving methods by following the model of other sciences that have achieved their power from the discovery of general principles. Clear examples from AI are the attempts to apply resolution, a uniform proof procedure for predicate calculus, to the development of general question-answering and problem-solving systems [e.g., Green (1969)]. In these efforts, questions or problems are translated into proposed theorems in the predicate calculus and a resolution theorem prover attempts to prove these theorems; the proof is then the basis for an answer or solution. The General Problem Solver (GPS) is an attempt to discover general methods that will solve a large class of problems without resorting to specific procedures.

The DENDRAL approach is a sharp contrast to these efforts. It was developed from the belief that since human experts have large quantities of detailed task-specific

Table 3-1 Organization of the DENDRAL programs

System	Components	Input	Output
Heuristic DENDRAL			
	Planning (Chap. 5):		
	MOLION	Mass spectrum	Molecular ion
	Planning Rule Generator	Planning rules	Constraints:
	PLANNER		superatoms
			GOODLIST
			BADLIST
	Generating (Chap. 4):		
	Acyclic generator		Candidate molecular structures
		Constraints	
	CONGEN		
	Testing (Chap. 6):		
	PREDICTOR	Candidate molecular structures	Most plausible structures
	MSPRUNE	Mass spec rules	Structures consistent with spectrum
Meta-DENDRAL (Chap. 7)			
	Planning:		
	INTSUM	Set of known structures and their mass spectra	Set of all possible fragmentations (ALLBRKS)
	Generating:		
	RULEGEN	Set of fragmentations	Candidate mass spec rules
	Testing:		
	RULEMOD	Candidate mass spec rules	Most plausible mass spec rules

knowledge, machines might profitably embody such information. While at some level of abstraction all problem solving and thought must have certain common features, we note that champion chess players are not invariably expert chemists or accomplished composers.

This observation does not mean that DENDRAL is of no interest beyond mass spectrometry. Generality, on this view, is to be sought at a higher level of abstraction, perhaps at the level of general procedures for knowledge acquisition. This theme will be explored more fully in the discussion of Meta-DENDRAL.

In large part, then, DENDRAL is a piece of "knowledge engineering," that is, the careful fashioning of many forms of knowledge into a complex system that functions smoothly.

3.4 OUTLINE OF DENDRAL PROGRAMS

Table 3-1 introduces the names of some of the component programs and data structures of the DENDRAL system and outlines their interrelationships.

The programs are written primarily in the list-processing language LISP [McCarthy (1960)] in the INTERLISP [Teitelman (1975)] implementation. Some routines are written in SAIL [VanLehn (1973)] and BCPL. At present the programs run on the Digital Equipment Corporation KI-10 computer at the SUMEX-AIM installation at Stanford University, Palo Alto, California. The BCPL programs have been exported to many other sites as well.

CHAPTER
FOUR

THE DENDRAL GENERATOR

The induction-by-enumeration model of scientific discovery was abandoned because of the seeming impossibility of enumerating all possible hypotheses. DENDRAL puts new life in the old model by proposing an enumerator of molecular structure hypotheses that is not only complete but nonredundant as well. This hypothesis generator can be guided by task-specific constraints in order to avoid exhaustive search of the hypothesis space. It is then no longer necessarily complete, but those hypotheses not generated are precisely characterized.

4.1 INTRODUCTION

The heart of the plan-generate-test paradigm is the generator. The seminal insight for DENDRAL was the original algorithm for exhaustively and nonredundantly generating acyclic structures as reported in Lederberg (1964b) and Lederberg (1965a). This algorithm, embodied in computer code,[1] was the basis of the first DENDRAL system. When the limitation to acyclic structures was overcome [Brown, Hjelmeland, and Masinter (1974), Brown and Masinter (1974)], DENDRAL increased its scope dramatically.

In this chapter we present an informal description of the cyclic generator, which includes the acylic generator as a component. The complete algorithm is complex. The correctness of the algorithm has been rigorously proved [Brown, Hjelmeland, and Masinter (1974)], but no procedure is known for proving that a program is a correct embodiment of a complex algorithm. However, the generator program has passed the important test of enumerating the correct *number* of structures for many cases where

---

[1] William C. White initially implemented Lederberg's algorithm in 1965 and Georgia Sutherland improved and expanded the scope of the program over the next several years.

the number of structural isomers was computed independently, and has been extensively checked against hand-calculated examples. We do not seriously doubt the program's correctness.

The chapter concludes with a description of a self-contained system called CONGEN that embodies the cyclic structure generator. Aside from its importance as a powerful system in its own right, CONGEN illustrates the basic concepts of CONstrained GENeration that underly the entire DENDRAL effort and the plan-generate-test paradigm.

4.2 OVERVIEW

The problem of enumerating all *ring-free* connectivity isomers from a set of atoms whose valences are known was solved with the construction of the original DENDRAL acyclic structure generator. The cyclic generator reduces the more general problem of enumerating *all* connectivity isomers to the special case of acyclic generation by treating ring structures as large atoms, called *ring superatoms*. A ring superatom is a chemical graph containing *one or more* connected rings and having *free valences*, that is, a number of unattached links are available for bonding to other atoms or ring superatoms. All atoms in a ring superatom are members of one or more rings. That is, there are no isthmuses—appendages that may be separated from the structure by cutting only one bond. Because in general a ring superatom is asymmetrical, its free valences, unlike those of an atom, may be distinguished from one another. To do so, with each ring superatom is associated a set of permutations on the nodes that specifies the symmetry of the structure.

Associated with any ring superatom is another graph, called a *vertex graph*, that has the same number of rings, connected in the same way, as the ring superatom. A vertex graph is an abstracted structure, and a number of ring superatoms correspond to the same vertex graph. Vertex graphs thus define equivalence classes of ring superatoms. A vertex graph together with a set of atoms (and their associated valences) underlies, and can be used to generate, all possible structurally equivalent ring superatoms for that set of atoms. Thus a set of vertex graphs together with atom/valence sets can generate sets of ring superatoms that may be combined with still other atoms by a tree generator to yield a set of connectivity isomers.

Atoms that are contained in more than one ring (i.e., those atoms at which rings are connected) are called *vertex atoms*. A single ring has no vertex atoms. Two rings sharing exactly one vertex atom are said to be *spiro-fused* (a chemistry term). More commonly, two rings share two vertex atoms, usually adjacent. The rings are said to be *edge-fused*, or, more commonly, *fused*. In defining ring superatom classes we ignore not only the atom names but also all nonvertex atoms. Thus a two-ring superatom connected with two vertex atoms is of the same vertex-graph form no matter whether the vertex atoms are adjacent or not, as illustrated in Figure 4-1.

When we consider only connectivity of ring superatoms vis à vis the vertex nodes we obtain *vertex graphs*. All nodes of these graphs have degree of at least 3; in fact the vertex graphs associated with most known chemical structures (except the limiting

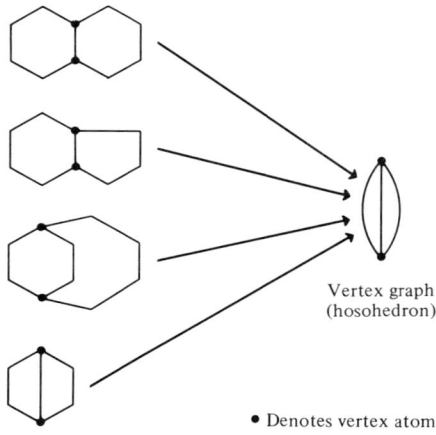

• Denotes vertex atom

Figure 4-1 Some of the two-ring superatoms that reduce to a common 2-trivalent-node vertex graph.

case of a single ring) have only nodes of degree 3. Spiro-fused ring structures have a node of degree 4, corresponding to the four links that fuse on a single vertex atom.

A CATALOG has been developed for most vertex graphs of chemical interest. The CATALOG as of this writing contains several thousand vertex graphs including all trivalent vertex graphs up to 18 nodes, plus some quadrivalent varieties. Higher-order graphs are handled by "tricks" that reduce them to trivalent and quadrivalent graphs. For example, a six-valent node may be considered to be a connection of two quadrivalent nodes:

$$\overset{\backslash\ /}{\underset{/\ \backslash}{-*-}} = \overset{\backslash}{\underset{/}{-*-}}\overset{/}{\underset{\backslash}{-*-}}$$

Some examples of vertex graphs and associated chemical structures are given in Figure 4-2.

Given an empirical formula, the enumeration of all isomers is a three-step process:

1. Ring generation
   a. *Form the initial partition*: Partition the atoms of the formula in all possible ways into sets, called *ring superatompots*, which will form ring superatoms, and a set, called the *remainingpot*, which will not be part of any ring superatom.
   b. *Identify vertex graphs*: For each partition, determine the (unique) vertex graph for each of the ring superatompots.
   c. *Generate equivalent ring superatoms*: Use each vertex graph to generate all ring superatoms in the equivalence class defined by that vertex graph.
2. Tree generation
   For each combination of ring superatoms and atoms from the remainingpot produced in step 1, use the acyclic generator to generate all possible tree structures. That is, treat ring superatoms as nodes in tree structures and generate all trees.
3. Expanding ring superatoms
   a. *Imbedding*: Expand each ring superatom node to its full structure in terms of atoms.
   b. *Pruning*: Throw out structures that would not be eliminated before imbedding.

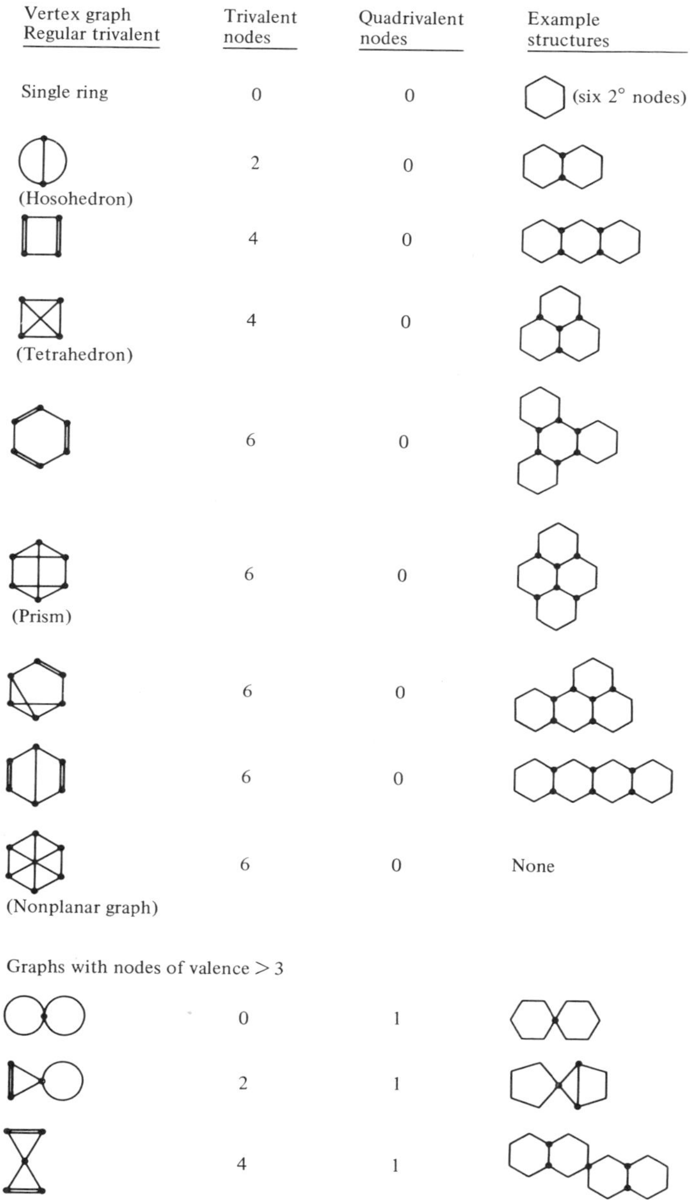

Vertex graph Regular trivalent	Trivalent nodes	Quadrivalent nodes	Example structures
Single ring	0	0	(six 2° nodes)
(Hosohedron)	2	0	
□	4	0	
⊠ (Tetrahedron)	4	0	
	6	0	
(Prism)	6	0	
	6	0	
	6	0	
(Nonplanar graph)	6	0	None
Graphs with nodes of valence > 3			
∞	0	1	
	2	1	
	4	1	

Figure 4-2 Some vertex graphs and associated chemical structures. (*Source:* Computer Representation and Manipulation of Chemical Information, *edited by W. T. Wipke, S. Heller, R. Feldmann and E. Hyde. Copyright 1974, John Wiley & Sons, Inc. Reproduced by permission. After Smith, Masinter, and Sridharan* [1974].)

4.3 RING GENERATION

4.3.1 Form the Initial Partition

The empirical formula is first converted to the abbreviated formula by computing degree of unsaturation, using Equation 1, page 12. The number of U's in the abbreviated formula determines how many ring superatoms are possible, since each ring superatom must have at least one unsaturation.[2] If there are three unsaturations available, they may form either one 3-ring superatom, or one 2-ring superatom plus a single ring, or three single rings. There is no known closed formula giving the number of partitions of a given number of U's, but their enumeration is straightforward.[3] For each partition of the unsaturations there are a number of ways to divide the multivalent atoms among each of the ring superatompots and the remainingpot. All univalent atoms (including the unmentioned H's) are assigned to the remainingpot, since they can never be included in rings.

Consider the example of $C_{10}H_{19}N$. The degree of unsaturation is 2, so the abbreviated formula is $C_{10}NU_2$. This means that two-ring superatoms are the most that need be considered. Remembering that it does not make sense for a ring superatompot to contain fewer than two atoms or less than one unsaturation, we have the following set of partitions.

The process next proceeds through several steps that will be illustrated for $C_{10}H_{19}N$, in Figure 4-3.

4.3.2 Identify Vertex Graphs

Construct the valence list. Let a_i denote the number of available atoms with valence i. The vector $(a_2, a_3, \ldots, a_n)$ is called the valence list for a ring superatompot containing atoms with n − 1 numerically distinct valences. For the ring superatompot containing C_9NU_2 the valence list is (0, 1, 9), because there are no atoms of valence 2, one of valence 3 (nitrogen), and nine atoms of valence 4 (carbon).

Compute the number of free valences. This computation is made by solving Equation 1, page 12, for free valence:

$$F = S - 2K + 2 - 2U \qquad \text{(Eq. 2)}$$

where S = sum of valences of multivalent atoms
  K = number of multivalent atoms
  U = degree of unsaturation (of a superatompot)

For the example, C_9NU_2, there are 17 free valences.

---

[2] The DENDRAL generator treats double bonds as rings and triple bonds as two-ring structures. See page 12.

[3] The number of partitions of n indistinguishable objects for n = 5, 10, and 20 is 7, 42, and 627, respectively. The number of partitions is asymptotic to $[4n(3)^{1/2}]^{-1} \exp[\pi(2n/3)^{1/2}]$.

Table 4-1 Possible partitions of $C_{10}NU_2$

Ring superatompot #1	Ring superatompot #2	Remainingpot	Ring superatompot #1	Ring superatompot #2	Remainingpot
$C_{10}NU_2$			C_3NU	C_5U	C_2
C_9NU_2		C	C_2NU	C_6U	C_2
C_8NU_2		C_2	CNU	C_7U	C_2
C_7NU_2		C_3	C_5NU	C_2U	C_3
C_6NU_2		C_4	C_4NU	C_3U	C_3
C_5NU_2		C_5	C_3NU	C_4U	C_3
C_4NU_2		C_6	C_2NU	C_5U	C_3
C_3NU_2		C_7	CNU	C_6U	C_3
C_2NU_2		C_8	C_4NU	C_2U	C_4
CNU_2		C_9	C_3NU	C_3U	C_4
$C_{10}U_2$		N	C_2NU	C_4U	C_4
C_9U_2		CN	CNU	C_5U	C_4
C_8U_2		C_2N	C_3NU	C_2U	C_5
C_7U_2		C_3N	C_2NU	C_3U	C_5
C_6U_2		C_4N	CNU	C_4U	C_5
C_5U_2		C_5N	C_2NU	C_2U	C_6
C_4U_2		C_6N	CNU	C_3U	C_6
C_3U_2		C_7N	CNU	C_2U	C_7
C_2U_2		C_8N	C_8U	C_2U	N
C_8NU	C_2U		C_7U	C_3U	N
C_7NU	C_3U		C_6U	C_4U	N
C_6NU	C_4U		C_5U	C_5U	N
C_5NU	C_5U		C_7U	C_2U	CN
C_4NU	C_6U		C_6U	C_3U	CN
C_3NU	C_7U		C_5U	C_4U	CN
C_2NU	C_8U		C_6U	C_2U	C_2N
CNU	C_9U		C_5U	C_3U	C_2N
C_7NU	C_2U	C	C_4U	C_4U	C_2N
C_6NU	C_3U	C	C_5U	C_2U	C_3N
C_5NU	C_4U	C	C_4U	C_3U	C_3N
C_4NU	C_5U	C	C_4U	C_2U	C_4N
C_3NU	C_6U	C	C_3U	C_3U	C_4N
C_2NU	C_7U	C	C_3U	C_2U	C_5N
CNU	C_8U	C	C_2U	C_2U	C_6N
C_6NU	C_2U	C_2			
C_5NU	C_3U	C_2			
C_4NU	C_4U	C_2			

Partition the free valences. The total number of free valences is divided among the atoms, none being assigned more than its valence less 2. In general this assignment can be done in more than one way, and each leads to a different class of possible structures. In our example we have 17 free valences to be assigned among one trivalent atom and nine quadrivalent atoms. If none is assigned to the trivalent, the 17 must

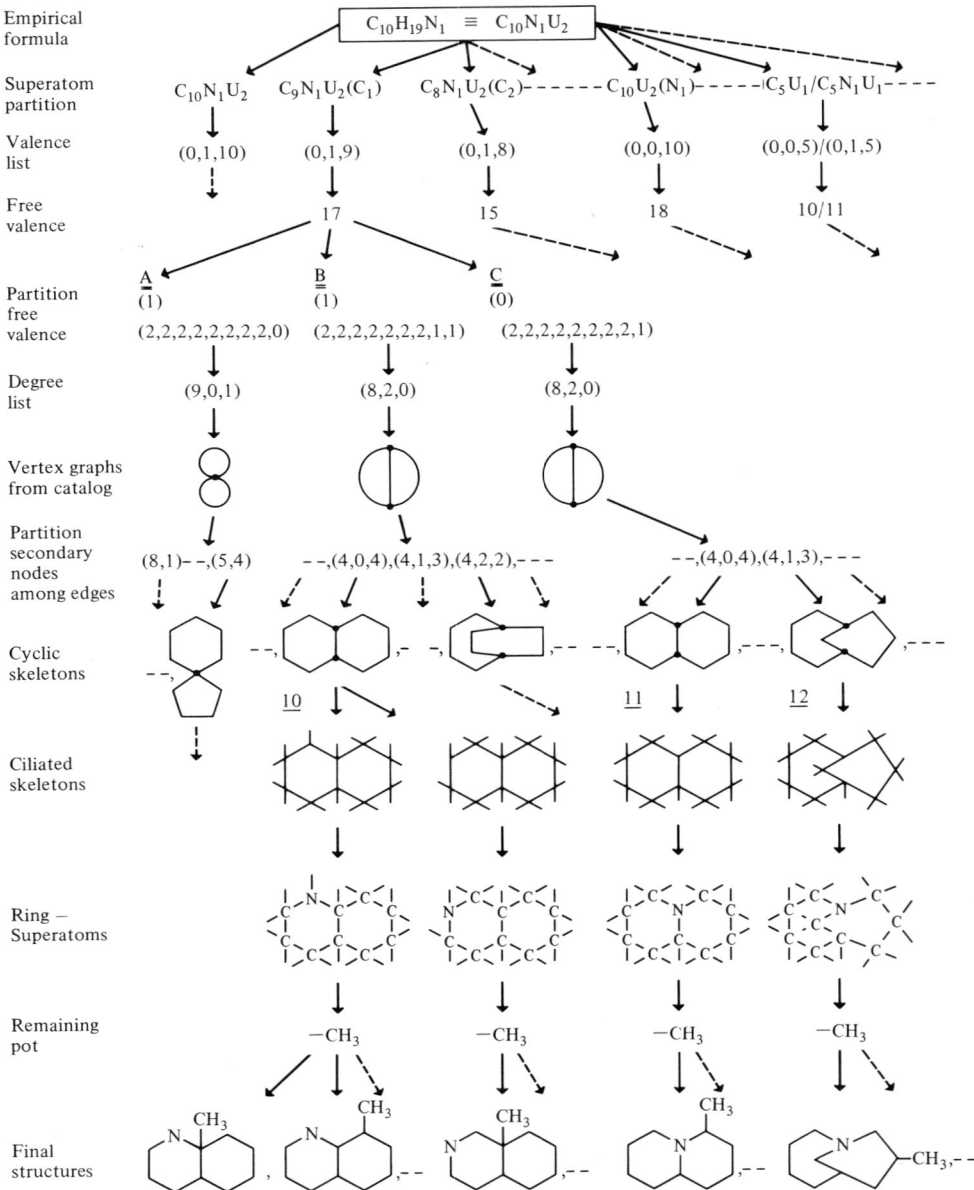

Figure 4-3 Structure generation for $C_{10}H_{19}N$. (*Source:* Computer Representation and Manipulation of Chemical Information, *edited by W. T. Wipke, S. Heller, R. Feldmann and E. Hyde. Copyright 1974, John Wiley & Sons, Inc. Reproduced by permission. After Smith, Masinter, and Sridharan [1974].*)

divide (2,2,2,2,2,2,2,2,1) among the nine quadrivalents. If one is assigned to the trivalent, there are two possible ways of assigning the remaining 16 among the 9 quadrivalents: (2,2,2,2,2,2,2,2,0) and (2,2,2,2,2,2,2,1,1).

Compute the degree list. The degree of an atom is simply its valence less the free valences assigned to it; for atoms in a ring superatom it is how many of its valences are used in forming the ring structure, since all such atoms by definition are members of one or more rings. Traditionally, atoms are said to be primary, secondary, tertiary, and quaternary for degree 1, 2, 3, and 4. Thus the carbon atom in CH_3- is primary, the carbon atom in $-CH_2-$ is secondary, and so forth. Let d_i be the number of atoms of degree i. The degree list is computed for a free-valence partition in the obvious way. Trivalent atoms with one free valence have degree 2, as do quadrivalent atoms with two free valences; trivalent atoms with no free valences and quadrivalent atoms with one free valence have degree 3, and so forth. Since atoms of degree 1 cannot be part of a ring superatom, the degree list $(d_2, d_3, \ldots, d_n)$ begins with the number of atoms of degree 2.

Determine what ring superatom organizations are possible. A degree list uniquely determines for its ring superatom the number of rings and their connectivity. In fact this structural information may be computed from the degree list by arithmetic operations. Once the degree list for a particular partition of a ring superatompot has been determined, it is "looked up" in the CATALOG.

The manipulations of the degree list that comprise the look-up are simple arithmetic operations that, interpreted in the graph-model semantics, amount to deleting recursively all secondary atoms one at a time until (1) no nodes remain (this is the special case of a single ring), or (2) only one node remains, with one or more reflexive links (these cases are called "daisies" and result from spiro-fusions and double bonds), or (3) two or more nodes remain, in which case the CATALOG is consulted for entries with this number of nodes.

4.3.3 Generate Equivalent Ring Superatoms

The remaining steps in the generation of ring superatoms from a given ring superatompot build on the selected vertex graph using the information contained in the associated free-valence partition and the degree list derived from it. The procedures involve, first, labeling the edges of the vertex graph with the number of atoms each will contain; second, labeling all nodes (the vertex nodes together with the just-generated nonvertex nodes) with degrees; and third, labeling each node with its particular atom type. Each step can, in general, be carried out in more than one way and all combinations must be considered.

Generate cyclic skeletons. The number of edges in the vertex graph is counted and the number of atoms of degree 2 is partitioned among these in all possible distinct ways, eliminating duplications due to the symmetry of the vertex graph. In the case of the hosohedron vertex graph (see Figure 4-2) associated with partition *B* in Figure 4-3, the three edges are equivalent because of the symmetry of the graph. The eight nodes of degree 2 may be assigned to the three edges in only 10 distinct ways: (8,0,0), (7,0,1), (6,0,2), (5,0,3), (4,0,4), (6,1,1), (5,1,2), (4,1,3), (4,2,2), (3,2,3). Each leads to a distinct cyclic skeleton with one node per atom. Note that after each labeling step, some of the symmetry of the unlabeled graph may be destroyed.

Assign degrees to nodes. The free valence partition is consulted to determine how many secondary nodes must have 0, 1, 2, etc., free valences; then how many tertiary nodes must have 0, 1, 2, etc., free valences; and so forth. Again there is in general more than one way to label the nodes subject to this constraint, while eliminating those that are equivalent by symmetry. This step yields *ciliated skeletons*.

Assign atoms to nodes. The final step is to assign the available atoms to the nodes that have been generated. All assignments consistent with the actual valences and node labelings must be made, again discounting structures that are equivalent by symmetry.

It is interesting to note that each of these three labeling steps is formally the same and in fact is carried out by the same labeling algorithm. Formally the problem is to associate a set of m labels, not necessarily distinct (in the three cases: secondary nodes, degrees, and atom names), with a set of n objects, not necessarily distinct (in the three cases: edges of a vertex graph, nodes of a cyclic skeleton, and nodes of a ciliated skeleton), to yield a set of labeled objects distinct up to a given set of symmetry transformations. We will not describe this algorithm in detail; it is published in Brown, Hjelmeland, and Masinter (1974).

4.4 TREE GENERATION—THE ACYCLIC GENERATOR

The ring superatoms generated by the above-described algorithm are next combined in all chemically possible ways with the atoms in the remainingpot. If we ignore the internal structure of the ring superatoms, the structures so created will be trees. The algorithm that accomplishes this structural creation is essentially the same as the original acyclic DENDRAL algorithm which enumerates all possible isomers for aliphatic structures. That algorithm will now be described, and the modification needed to incorporate it into the general cyclic structure generator will be described later.

4.4.1 DENDRAL Notation

It is convenient to have a linear notation for representing tree structures, both because for some purposes tree-structure diagrams are unwieldy, and because a linear notation frequently makes enumeration and formal manipulation more uniform and systematic. The DENDRAL notation is a linear notation for aliphatic chemical structures.[4] Other chemical nomenclatures have been devised for similar purposes, but the DENDRAL notation is more systematic and perspicuous.

---

[4] A linear notation was devised for cyclic structures, but we were unable to map it into a generating algorithm in the same way as for acyclic structures.

The DENDRAL notation is based on a method for finding a unique ordering, i.e., a unique path through the nodes and branches of any given acyclic structure diagram. This method is also the basis for enumerating all aliphatic connectivity isomers of a given empirical formula because the notational algorithm can be easily turned into a generating algorithm, as discussed here. The method could be extended to an enumeration algorithm for all acyclic structures by proceeding through all (countably infinite) empirical formulas in the obvious dictionary order.

It is apparent that we must have a way of identifying a unique and characteristic starting point for traversing the tree. Otherwise, the same tree would have several possible encodings, and it could become a monumental task to determine from two codes the relation between the trees they represent. In particular, it would be difficult or impossible to know if two codes represented the same tree. Such a situation would make an exhaustive, nonredundant enumeration algorithm inefficient or impossible.

Fortunately, for the case of trees, there is always a unique starting point, the *centroid*. This was proved by the mathematician Jordan in 1869 [see reference 14 of Buchanan, Duffield, and Robertson (1971)]. There are three cases to consider: (A1) For trees with an odd number of nodes, the centroid is that unique node (*A*tom) from which each branch carries less than half the total number of nodes. (A2) For trees with an even number of nodes, the centroid is that node (*A*tom) from which each branch carries less than half the total number of nodes, *if such a node exists*; otherwise, (B) the centroid is an edge (*B*ond) that joins nodes of equal node count.

The construction of the unique, unambiguous DENDRAL notation for any given tree is now straightforward; we need only establish conventions for deciding which route to take when several are available. These conventions are called the *DENDRAL Canons of Order*. Once these have been adopted, it remains only to convert the thus-defined canonical traversal of a tree into a linear notation.

The most familiar and direct method for uniquely encoding tree structures in linear format is the use of parentheses in the usual way: each terminal node is enclosed in parentheses, and subnodes of a single parent are separated (say by commas) and enclosed in parentheses. The procedure is iterated until the top node is reached (note that 'terminal node', 'top', 'subnodes', and 'parent' were not uniquely defined until a topmost node was specified). For complex trees a more compact form is Polish prefix notation; for chemical graphs we may use . and : as the (unary) operators corresponding to single and double bonds.[5] Propane would be represented as $(CH_2((CH_3)(CH_3)))$ in parenthetic notation or as $CH_2 . . CH_3CH_3$ in DENDRAL-like notation.[6] Here we have collapsed all the hydrogens onto their associated carbon nodes to simplify the notation. Since the number of hydrogens is readily determined for each carbon (4 minus the number of other links), we can and do omit mention of hydrogens. The DENDRAL notation for propane then becomes C . . CC.

---

[5] For enumeration of aliphatic compounds bonds may be single, double, or triple. In the context of the cyclic generator all bonds are the same type since doubly and triply bonded atoms are treated as ring superatoms.

[6] The parallel between this representation of chemical structures and the LISP language's representation of branching trees (lists) was one reason why LISP was selected initially for the implementation of DENDRAL.

4.4.2 Enumeration of Hydrocarbons

We begin with the special case of the hydrocarbons (all atoms are carbon or hydrogen) for which the canons of order reduce to a small set. As usual, hydrogen atoms are ignored until the very end. Enumeration begins by choosing a possible centroid, then partitioning the number of carbon atoms in the formula in all possible ways between the atoms connected by the centroid (if it is a bond) or among the bonds issuing from the centroid (if it is an atom). Our first canons of order fix the order in which centroids are considered. Canon 1: consider bond centroids before atom centroids; Canon 2: consider lower-degree atoms before higher-degree atoms in generating atom centroids. (Recall that a first-degree atom is an atom with one nonhydrogen bond, a second-degree atom has two nonhydrogen bonds, and so forth.) Clearly a first-degree node can never be a centroid, and a second-degree node cannot be a centroid if the structure has an even-node count.

In the consideration of a bond as a centroid, there is only one division of the nodes: half and half. This division is also true for a centroid that is a second-degree node. For third- and fourth-degree nodes there are in general several ways of partition-

Table 4-2 DENDRAL enumeration of the pentyl radicals

Number	Graph	DENDRAL formula
1	—C—C—C—C—C	.C.C.C.C.C
2	—C—C—C—C with C branch	.C.C.C..CC
3	—C—C—C—C with C branch	.C.C..CC.C
4	—C—C—C with two C branches	.C.C...CCC
5	—C—C—C—C with C branch	.C..CC.C.C
6	—C—C—C with two C branches	.C..CC..CC
7	—C—C—C with C—C branch	.C..C.CC.C
8	—C—C—C with C, C branches	.C...CCC.C

ing the remaining nodes in such a way that each component of the partition contains less than half the total. A partition is a set of integers; choose as a canonical notation an ordering of these integers (a vector) into a nondecreasing sequence. The vectors have a natural ordering: $(a_i) < (b_i)$, $i = 1$ to n if and only if for some $k < n + 1$, $a_k < b_k$ and $a_j = b_j$ for $1 < j < k$. Canon 3: consider the partitions in ascending sequence. This means that radicals of lower-node count will precede those of higher-node count in the canonical DENDRAL notation for a particular structure, and that structures with lower-node counts will be enumerated sooner by the acyclic generator.

Given a centroid and a partition, we must consider for each component of the partition all possible radicals that may be constructed from that set of atoms. These must be enumerated in a systematic order.

In enumerating the radicals for a given set of atoms, the starting point is uniquely determined; it is the site of the free valence. This is called the *apical node*. Canon 4: consider all possible degrees of the apical node in ascending order. Then consider all partitions of the remaining atoms among the branches from the apical node, remembering that there is no "balance" constraint now since the apical node is not a centroid. Canon 5: These partitions are ordered as before (Canon 3).

For each component of a partition, apply the procedure, just described, recursively. This algorithm yields, for the case of five carbons (pentyl radicals), the enumeration in Table 4-2.

Now we will illustrate the enumeration of a class of hydrocarbons, the hexanes. The entire algorithm yields the enumeration depicted in Figure 4-4. The node count is even (6), so we consider the bond centroid possibility first. This yields the single partition (3,3). The enumeration of the propyl radicals by the above algorithm yields only two possibilities: .C.C.C and .C..CC in that order. Pairwise enumeration of the propyl radicals gives (.C.C.C and .C.C.C), then (.C.C.C and .C..CC), then (.C..CC and .C..CC).

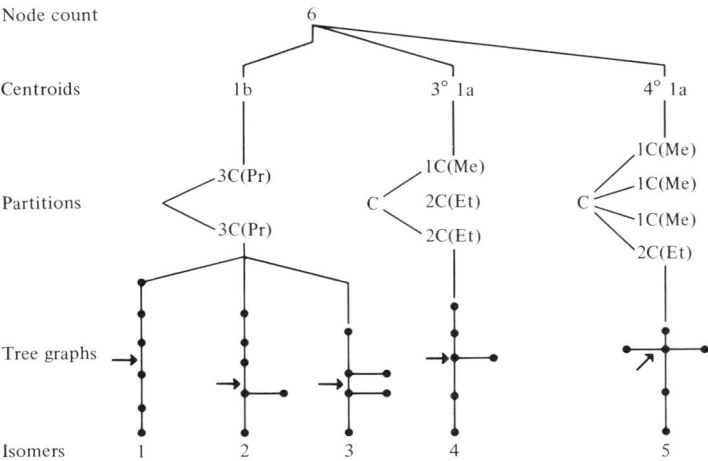

Figure 4-4 Dendral enumeration of the hexanes, with centroids marked by arrows. (*Source:* Mass Spectrometry: Techniques and Applications, *edited by G. W. A. Milne, Copyright 1971, John Wiley & Sons, Inc. Reproduced by permission. After Buchanan, Duffield, and Robertson [1971].*)

Table 4-3 The number of isomers of some simple radicals and saturated hydrocarbons

	Number of carbons	Number of isomers
Radicals		
Methyl	1	1
Ethyl	2	1
Propyl	3	2
Butyl	4	4
Pentyl	5	8
Alkanes		
Methane	1	1
Ethane	2	1
Propane	3	1
Butane	4	2
Pentane	5	3
Hexane	6	5
Heptane	7	9
Octane	8	18
Nonane	9	35
Decane	10	75
Isocane	20	366319

Thus we have three isomers with a bond centroid. There can be no first-degree atom centroids for *any* structure, and no second-degree atom centroids in this (even-node count) structure. Choosing a third-degree atom centroid allows only one partition of the remaining five atoms (remember that no component of the partition can be more than half the total), yielding only one isomer. This is also the case for the last enumerated isomer resulting from a fourth-degree atom centroid.

Table 4-3 is provided for those readers who wish to test their understanding of the algorithm by trying some other examples. (Ten atoms challenge almost everyone's ingenuity.)

4.4.3 Enumerating Other Acyclic Structures

For compounds containing heteroatoms (nonhydrogen, noncarbon atom) we need another canon for choosing order. Canon 6: order atoms (arbitrarily) by increasing atomic number: C, N, O, P, S for these five common atoms. (The order of these most common atoms is easily memorized, since it is also alphabetical.) Whenever one must sort a set of atoms, it is done on atomic number.

The final decision concerns unsaturation: what to do when structures contain double and triple bonds. Canon 7: consider single bonds first, then double bonds, then triple bonds.

4.5 THE CYCLIC GENERATOR

For a given set of ring superatoms (one exemplar for each ring superatompot generated by the cyclic structure generator) and the atoms of the remainingpot, we can apply a modified version of the acyclic generator to enumerate all possible single structures. The modifications to the generator are necessary because the free valences of a ring superatom, unlike those of an atom, are distinguishable; this modification is readily handled by supplying the symmetry transformations of each ring superatom to the acyclic generator, which can then reject duplications. A final addition to our canons of order is also needed because it is necessary to have a rule for ordering superatoms.

We now have a complete set of canons of order that uniquely determines the order of enumeration of all structures for compounds of C, H, N, O, P, and S:

1. Canon 1: Consider bond centroids before atom centroids.
2. Canon 2: Consider lower-degree atom centroids before candidates of higher degree.
3. Canon 3: Consider partitions of atoms in their natural order.
4. Canon 4: In generating radicals, consider possible degrees of the apical node in ascending order.
5. Canon 5: In generating radicals, consider partitions of atoms in their natural order.
6. Canon 6: Consider atoms in the order C, N, O, P, S.
7. Canon 7: Consider single bonds before double bonds before triple bonds.
8. Canon 8: Consider smaller ring superatoms before larger, and for those of equal number of atoms list their constituent atoms in order of atomic number and order the ring superatoms alphabetically on this naming rule.

Together, the rules for partitioning atoms into superatompots, plus the rules for enumerating ring structures for a given superatompot, plus the acyclic canons of order yield a complete, exhaustive, nonredundant cyclic structure generator.

4.6 CONSTRAINING THE GENERATOR: CONGEN

The cyclic generator is embodied in the full DENDRAL plan-generate-test system. In subsequent chapters the planner and tester will be described. There it will be seen how a complete structure elucidation engine can be realized. However, the cyclic generator alone is a very powerful program. In the hands of chemists it can prove a useful adjunct to their own planning and testing heuristics, especially when they can specify constraints on the structures in terms of chemical graphs. Such a program has been implemented. It is called the CONstrained GENerator, or CONGEN.

We will now describe CONGEN because it demonstrates in clear fashion the concept of heuristic search in the space of possible chemical structures. It is, incidentally, a well-documented and user-oriented program with a user's manual[7] and online help

---

[7] Available at cost from The DENDRAL Project, Computer Science Department, Stanford University, Stanford, California 94305.

facilities. A serious effort has been made to "export" it to the scientific community. Thus it is of interest in its own right.

For problems of serious concern to the chemist, the number of connectivity isomers for a given empirical formula is so large as to be unmanageable (see Section 8.2). Of course one normally knows much more about a compound than its empirical formula. What is known comes from many sources, a mass spectrum often being an important one. The data can often be interpreted in a form that rules out certain substructures and strongly suggests the presence of others. By instructing the generator to generate only structures containing the required substructures and not containing forbidden substructures, the number of generated isomers may be dramatically reduced.

CONGEN permits the user to constrain the enumeration by specifying several types of constraints.[8] These are:

1. SUBSTRUCTURE constraints: These are substructures that must be present with specific cardinality. The specific number (including none) of occurrences may be specified, or a range for the minimum and maximum number of occurrences may be given. (Superatoms specified by the chemist with a definite number of occurrences may be used from the start in place of their constituent atoms.)
2. RING constraints: The user may specify *sizes* of rings that must be present. The number of occurrences (including none) or a range for the number of occurrences of a given ring size may be specified.
3. PROTON constraints: The user may specify the number of hydrogen atoms that must be associated with a given structure, without being specific as to where they are bonded.
4. ISOPRENE constraints: The user may specify a range for the number of isoprene units that must be part of the generated structures. An isoprene unit is a Y-shaped grouping of five carbon atoms.

The last mentioned constraint is included because many naturally occurring compounds contain known numbers of isoprene units. The SUBSTRUCTURE constraint mechanism is sufficiently general to encompass ISOPRENE constraints, but because they are an important and ubiquitous special case, they are treated separately. The strategy is sufficiently flexible to permit further extensions along these lines, and time will no doubt see further special cases handled.

The first step in using CONGEN is to define *superatoms*. We have encountered the concept of ring superatoms above. A superatom may be either a ring superatom or an acyclic structure containing more than one atom. That is, it is any connected graph that is treated as a unit, acting in the role of an atom in the construction of larger structures.

The specification of superatoms is done with DEFINE commands. These permit the construction of computer representations of rings and chains of atoms of specified type and the assignment of free valences. Certain atoms may be TAGged to specify

---

[8] At present there is no way to specify constraints on the three-dimensional configuration of structures, although this problem is a focus of current work. See Section 4.6.2.

PROTON constraints. Ranges of hydrogen attachment may also be specified for each atom.

A set of constraints of the above listed types may now be specified. This set of constraints applies to the first step of generation, which treats the specified superatoms as units distinguished only by their free valences. This is the GENERATE step, which produces all possible structures of atoms and superatoms following the algorithm of the cyclic generator but rejecting all structures violating defined constraints.

The final step, called IMBED, expands ("explodes") the superatoms one type at a time so that the resulting structures are representations of chemical structures in terms of atoms and bonds only. During imbedding, structures are also constrained according to the user's specifications, since the expansion of a superatom could give rise to violations of constraints that were not violated by intermediate structures. For example, an excluded substructure that consists of portions from each of two permitted substructures might emerge as these substructures are expanded when they are connected in one way, but not when they are connected in other ways. The set of constraints may differ for each IMBED step.

The program is interactive, permitting chemists to revise their list of structures and constraints at any stage. Thus if the generation produces an inordinate number of possibilities, they may add further constraint information until they are able to winnow the set down to manageable size. The final enumeration ideally would be a single structure, though in practice this is not often the case. However, if the number of structures is sufficiently small to permit examination of each member, the chemist's stock of judgment and intuition that remains uncodified may, at times, be sufficient to make a good guess as to the correct structure.

4.6.1 An Example of CONGEN Use

The following is a record of a session with CONGEN. The example is very simple and of no chemical interest but serves to illustrate the program's syntax and behavior.

The chemical information available is as follows:

C1. The empirical formula is $C_{12}H_{14}O$.
C2. The compound contains a keto group in a five-membered ring.
C3. There are three protons (H's) alpha to the carbonyl group.
C4. There are two vinyl groups (—C=C—) and (C4B) four vinyl protons.
C5. There is no conjugation (alternation of double and single bonds).
C6. There are no diallylic protons (hydrogens at the middle carbon of a diallylic structure: —C=C—C—C=C—), (C6B) nor protons alpha to both a vinyl and the keto group.
C7. There are only two quaternary carbons, one in the keto group and one in one of the vinyl groups.
C8. There are no additional multiple bonds.
C9. It is assumed there are no three- or four-membered rings.
C10. There are no methyl groups.

The following typescript of a dialogue with CONGEN shows how the complete set of topological isomers satisfying these constraints can be obtained. Fitting the constraints to the CONGEN syntax sometimes requires some roundabout procedures. The program is undergoing constant revision, in part to make these communications more succinct and natural to the chemist. Our example illustrates the system as of this writing.

Comments written in lowercase and enclosed in brackets have been inserted into the transcript to point out the purpose of each section with respect to the above constraints. The CONGEN syntax should be decipherable with the following brief guide.

CONGEN prompts the user with #. The main subprograms that may be called when this prompt is given are DEFINE, GENERATE, and IMBED. Each of these subprograms issues a > prompt to which the user replies with subcommands. User-typed responses are in bold type.

Substructures are defined with the DEFINE commands. Some of the substructures so defined will be used to specify superatoms that are to occur in the final graphs, but others will be disallowed. The DEFINE commands

RING n

and

CHAIN n

define a ring and a chain, respectively, of *n carbon* atoms and assign to each atom an identification number. To define superatoms with other atom types, RING and CHAIN are followed by the ATNAME command, which changes atom types. For example,

ATNAME 3 N 5 0

would make atom 3 a nitrogen and atom 5 an oxygen; the argument list can have as many number-name pairs as desired. More complex superatoms are defined by modifying rings and chains with the

LINK a b m

command, which connects atom *a* to atom *b* with a chain of *m* carbons. This is also how multiple bonds may be specified;

LINK 1 1 1

doubly bonds a carbon to atom 1.

ATOMFV a v

is used to define the (exact) number of free valences for specified atoms; the argument list is an arbitrarily long list of atom-number/free-valence-count pairs.

HRANGE a n m

specifies that the number of hydrogens bonded to atom *a* must be between *n* and *m*, inclusive.

More complex structures may be defined by naming superatoms within superatoms, using ATNAME to replace atom names in the new superatom with names of previously defined superatoms.

DEFINE has commands for diagramming superatoms, showing atoms by number (DRAW NUMBERED) or by type (DRAW ATNAMED), or displaying them in tabular form (SHOW). These features are illustrated in the example.

GENERATE is the command that produces all intermediate structures meeting the defined constraints, with the superatoms not imbedded ("exploded"), but treated as single atoms. After GENERATE is called, it gives the user the opportunity to define the CONSTRAINTS. After generating, more structures may be defined to give new constraints for the IMBED steps. IMBEDding is done one superatom at a time, allowing yet more flexibility in application of constraints. After GENERATE and IMBED, the user may DRAWSOME of the structures.

Sample CONGEN Dialogue

[Comments are bracketed and refer to problem definition statements C1 to C10, above.]

WELCOME TO CONGEN, VERSION VI.
#DEFINE MOLFORM C 12 H 14 O
MOLECULAR FORMULA DEFINED

#DEFINE SUBSTRUCTURE Z   [Z is the structure required by C2 and C3.]

(NEW SUBSTRUCTURE)
>RING 5   [C2: form five-membered ring.]
>LINK 1 1 1   [C2: add keto group.]
>ATNAME 6 O
>HRANGE 2 1 1 3 1 2 4 1 2 5 2 2

                                        [C3: atom 5 will have 2 H's, which will be alpha to the carbonyl; atom 2 will have the third H alpha to C=O.]

>DRAW NUMBERED

SUBSTRUCTURE Z (HRANGES NOT INDICATED)
NON-C ATOMS: 6->O

```
         /
 6=1-5-4
  \  |
   2-3
   |  \
```

58 APPLICATIONS OF ARTIFICIAL INTELLIGENCE FOR ORGANIC CHEMISTRY

>SHOW

SUBSTRUCTURE Z

ATOM#	TYPE	NEIGHBORS				HRANGE
1	C	6	6	5	2	
2	C	3	1	FV		
3	C	4	2	FV		1-2
4	C	5	3	FV		1-2
5	C	1	4			
6	O	1	1			

>DONE
Z DEFINED

#DEFINE SUBSTRUCTURE V   [V is the vinyl group.]
(NEW SUBSTRUCTURE)
>RING 2
>DRAW ATNAMED

SUBSTRUCTURE V:

C=C

>DONE
V DEFINED

[Comment: Next the disallowed structures are defined.]

#DEFINE SUBSTRUCTURE VV   [C5: VV is a double vinyl, which would
                                    violate conjugation prohibition.]
(NEW SUBSTRUCTURE)
>CHAIN 2
>ATNAME 1 V 2 V
>DRAW ATNAMED

SUBSTRUCTURE VV:

V-V

>DONE
VV DEFINED

#DEFINE SUBSTRUCTURE VCHV   [C6: VCHV is a diallyl with 1 or 2 H's,
                                    which is prohibited.]

(NEW SUBSTRUCTURE)
>**CHAIN 3**
>**ATNAME 1 V 3 V**
>**HRANGE 2 1 2**
>**DRAW ATNAMED**

SUBSTRUCTURE VCHV: (HRANGES NOT INDICATED)

V-C-V

>**DONE**
VCHV DEFINED

#**DEFINE SUBSTRUCTURE CH0**   [C7: CH0 is a quaternary carbon.]
(NEW SUBSTRUCTURE)
>**CHAIN 1**
>**HRANGE 1 0 0**
>**DONE**
CH0 DEFINED

#**DEFINE SUBSTRUCTURE CH3**   [C10: CH3 is a methyl group.]
(NEW SUBSTRUCTURE)
>**CHAIN 1**
>**HRANGE 1 3 3**
>**DONE**
CH3 DEFINED

#**DEFINE SUBSTRUCTURE VH**   [C4B: "tagged" vinyls needed to specify
                                       PROTON constraint later.]

(NEW SUBSTRUCTURE)
>**CHAIN 1**
>**ATNAME 1 V**
>**TAG 1**
>**SHOW**

SUBSTRUCTURE VH:

ATOM# TYPE NEIGHBORS
  1*   V
TAGGED ATOMS ARE INDICATED BY A *

>**DONE**
VH DEFINED

#GENERATE
SUPERATOM: Z
RANGE OF OCCURRENCES: AT LEAST 1
SUPERATOM: V
RANGE OF OCCURRENCES: AT LEAST 2
SUPERATOM:
'COLLAPSED' FORMULA IS C 3 Z 1 V 2 H 9
CONSTRAINT: **LOOP Z NONE**   [This constraint prevents Z from bonding with itself.]
CONSTRAINT: **SUBSTRUCTURE CH3 NONE**
     [C10]
CONSTRAINT: **SUBSTRUCTURE CH0 NONE**
     [C7]
CONSTRAINT: **SUBSTRUCTURE VV NONE**
     [C5]
CONSTRAINT: **SUBSTRUCTURE VCHV NONE**
     [C6]
CONSTRAINT: **RING 3 NONE**   [C9]
CONSTRAINT: **RING 4 NONE**   [C9]
CONSTRAINT: **RING 2 EXACTLY 3**   [There are already 3 double bonds (carbonyl plus two vinyls). This constraint makes certain there are no more and no triple bonds.]
CONSTRAINT: **PROTON VH EXACTLY 4**
     [C4B]
CONSTRAINT:

[Comment: Each time CONGEN generates a structure it prints ".".]

. . . .

[Comment: Now first-level intermediate structures have been generated.]

18 STRUCTURES WERE GENERATED

[Comment: The following is one of the 18 structures; as required, it contains one Z and two V's that will be imbedded later. Most of the constraints have been employed already, but a few remain for the imbedding steps.]

#DRAW ATNAMED 1

#1:

```
C-C-V
| | |
V-C-Z
```

THE DENDRAL GENERATOR

#DEFINE SUBSTRUCTURE VCHCO   [C6B: VCHCO defines protons alpha to both V and C=O to be prohibited.]

(NEW SUBSTRUCTURE)
>CHAIN 4
>JOIN 3 4
>ATNAME 4 O
>ATNAME 1 V
>DRAW ATNAMED

SUBSTRUCTURE VCHCO:

V-C-C=O

>HRANGE 2 1 2
>DONE
VCHCO DEFINED

#IMBED

[Comment: Z will now be imbedded first. More superatoms could be imbedded now also.]

SUPERATOM: Z
NUMBER TO BE IMBEDDED:1
SUPERATOM:
THE 'EXPANDED' FORMULA IS O 1 C 8 V 2 H 14
CONSTRAINT:SUBSTRUCTURE VCHCO NONE
                                                      [C6B]
CONSTRAINT:RING 3 NONE   [C9]
CONSTRAINT:RING 4 NONE   [C9]
CONSTRAINT:

[Comment: As CONGEN begins to consider each structure it prints "#", and for each new structure produced it prints ".".]

#..#..#.#.#....#....##..######....#....#..##..
29 STRUCTURES WERE OBTAINED
#DRAW ATNAMED 1

[Comment: The following is a sample structure in which Z has been imbedded. Note that the V's are not yet imbedded.]

#1:

```
 0
 =
  C         V
 / \       / \
C   C---C   C
 \  |    \ /
  C-C     C
     \   /
      \ /
       V
```

#IMBED

[Comment: The V's will now be imbedded.]

SUPERATOM: V
NUMBER TO BE IMBEDDED: 2
SUPERATOM:
THE 'EXPANDED' FORMULA IS O 1 C 12 H 14
CONSTRAINT: **SUBSTRUCTURE CH0 EXACTLY 2**

                                  [C7: we must end up with exactly two quaternary carbons.]

CONSTRAINT: **RING 3 NONE**   [C9]
CONSTRAINT: **RING 4 NONE**   [C9]
CONSTRAINT:
#.#.#.#..#..#.#..#..#..#..#..#..#.#.#.#..#.#..
#.#..#.#..#..#...#..#..#.#.#.#..
47 STRUCTURES WERE OBTAINED

#DRAW ATNAMED (4 6)

[Comment: The following is a selection of final structures.]

#4:

```
      C--C
     /|  =\
    C C  C \
    | |  |  C
    | |  |  |
    C-C--C--C-C=C
    =
    0
```

THE DENDRAL GENERATOR 63

#5:

```
 O           C
 =           =
 C-C         C
 | |\       /
 | | C---C
 C C |   |
  \| |   |
   C-C   C
    = /
     C
```

#6:

```
              C
              =
      C---C-C
     /    |  \
O==C      |   C C
    \     | / =|
     \    C   ≠
      \ /     |=
       C      | C
        \     |/
         C---C
```

#DRAW ATNAMED 30 31

#30:

```
           C
         / =
        /   C
   C--C   /|
   |   \ C |
   |    X  |
   C---C-C-C--C=C
   =
   0
```

#31:

```
       C
      / \
     /   \
    C  C===C
   /|/     |
  C C      |
   \ \     |
    C-C---C--C--C==C
    =
    0
```

#EXIT
DO YOU WANT TO SAVE YOUR SESSION ON FILE?: YES
FILE NAME: CONGEN.EXAMPLE
SAVED ON CONGEN.EXAMPLE
EXIT

4.6.2 Stereoisomer Generation

After the connectivity isomers have been generated, it is possible to elaborate some of the three-dimensional properties of those structures or examine the list in various ways. CONGEN has deliberately been kept unencumbered by these options. Additional programs have been written to allow users to look at the different relative orientations of groups in space (STEREO), rank or remove structures with respect to arbitrary assignments of scores to features (EXAMINE), and determine the plausibility of structures by looking at known reaction products and the likelihood that each structure would produce those products (REACT). These and other programs are currently under development and so will be mentioned only briefly.

4.6.3 STEREO

Much of organic chemistry involves the positioning of atoms in space. CONGEN is only a first step toward providing final answers to many kinds of problems. For example, the biological activity of drugs is known to be closely tied to their three-dimensional "fitting" into receptor molecules in the body.

Within the large problem of modeling the positions of atoms in space, it is important to specify the possible orientations of parts of a molecule around one or more appropriately substituted trivalent and tetravalent atoms, called stereocenters. The STEREO program[9] looks at one structure from CONGEN at a time and defines the complete set of alternative orientations of parts around all stereocenters, thus producing for each topological structure in the CONGEN list a set of stereoisomers. For ex-

---

[9] The STEREO program was designed and implemented by Dr. James Nourse, who also developed its theoretical foundations in graph theory and group theory [Nourse (1978b) and Nourse (1979a)].

ample, the central carbon in the structure below is a stereocenter in which the four substituents (S1 to S4) are positioned on the vertices of an imaginary tetrahedron enclosing the central carbon. The four substituents (S1 to S4) can be positioned either above or below the plane of the paper:

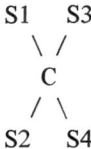

Different stereoisomers result from the following positionings:

Above the plane	Below the plane
S1, S4	S2, S3
S2, S3	S1, S4

When there are n stereocenters, there are at most 2^n distinct stereoisomers. Two important features of the generator parallel features of CONGEN: (1) it produces only the subset of the 2^n potential stereoisomers that *are* distinct and (2) it can be constrained to avoid (prospectively) classes of stereoisomers that are less plausible, for one reason or another.

4.6.4 EXAMINE

Even though a chemist has defined required and forbidden substructures as part of the specification to CONGEN, it is useful to provide additional interactive capabilities for examining the list of CONGEN structures under different hypotheses. The EXAMINE program[10] allows the user to define additional structural features and find subsets of structures from the CONGEN list that satisfy Boolean combinations of features.

It is useful to be able to ask how many structures in a CONGEN list contain isoprene units, spiro forms, ring fusions, and complex functional groups without having to inspect a drawing of each one. Beyond this function, the program allows the chemist to mark the structures found, draw them, save them, remove them, rank them, or replace the original CONGEN list with the list of structures found.

An example is shown below.[11] The program's requests and messages are printed in capitals; the user's typing is in bold type and is lowercase.

INPUT FILE: **azocine-saved.32st;1** [OLD VERSION]
READING <GRAY>AZOCINE-SAVED.32ST;1
THIS IS A FILE WRITTEN BY CONGEN

.
.
.

---

[10] EXAMINE was implemented by Dr. Neil Gray.
[11] From pp. 3–4 of STRUCC manual, by N. A. B. Gray, Stanford University, August 1978.

(32 STRUCTURES)
  xmn
DO YOU REQUIRE SIMPLY TO PRUNE YOUR STRUCTURE LIST?: n
DO YOU WANT TO RANK YOUR STRUCTURES? n
DO YOU WANT TO USE A LIBRARY? y
FILE NAME: <smith>examine . library;1 [OLD VERSION]
READING <SMITH>EXAMINE . LIBRARY;1
DO YOU WANT ALL SUBSTRUCTURES IN THE FILE? y
  (FILE READ OK)
     .
     .
     .
SUBSTRUCTURE C=C ... PRESENT IN 32 STRUCTURES.
SUBSTRUCTURE MONOSUBSC=C ... PRESENT IN 19 STRUCTURES.
     .
     .
     .
SUBSTRUCTURE ALKYNE ... PRESENT IN 4 STRUCTURES.
SUBSTRUCTURE ALLENE ... PRESENT IN 0 STRUCTURES.
     .
     .
     .
SUBSTRUCTURE NITRILE ... PRESENT IN 0 STRUCTURES.
ENTER COMMANDS FOR SELECTING SUBSETS OF STRUCTURES WITH PARTICULAR FEATURES.
32 STRUCTURES
-> select
> alkyne
4 STRUCTURES WITH ALKYNE
-> remove
28 STRUCTURES WITH (NOT ALKYNE)
-> done

DO YOU WANT TO CHECK FOR OTHER FEATURES?: n
DO YOU WANT TO FILE ANY OF YOUR SELECTION FEATURES?: n
WOULD YOU LIKE TO KEEP THE SUBSTRUCTURES READ FROM THE FILE? n

4.6.5 REACT

Important structural information about an unknown compound comes from knowing what chemical reactions produced the compound or what products are formed in reactions with it. The REACT program [Varkony et al. (1978a) and Varkony et al. (1978b)] simulates chemical reactions to use knowledge of reactions in structure determination problems. The chemist interactively defines chemical reactions for the program to use. These reactions are carried out by the program on each of the candidate structures in the CONGEN list to show the chemist what products one would be likely to observe from each of the possible candidates. If some products are known, then the CONGEN list can be pruned manually to delete candidates that do not lead to those products.

Another use of REACT is to delineate the structures that could arise from known precursors through a complex chain of reactions. The likely products, then, constitute the most plausible subset of the CONGEN list of candidates for a specific problem. For example, REACT has been applied to known sterol structures to help determine which of the CONGEN possibilities would be likely to be formed in nature through known biosynthetic reactions and chains of known reactions. It is difficult to follow complex chains manually in order to explore the possibilities. REACT is able to follow the reaction pathways from the known marine and plant sterols (combinations of seven naturally occuring sterol skeletons with over 30 known alkyl side chains on the C-17 position), using nine different reactions and sequences of them. As with the other programs, REACT is used with chemical constraints on the production of the tree of possible pathways and products.

In the case of the sterols, the total number of sterols (of many empirical formulas) that can be produced from the naturally occurring skeletons and side chains (within biosynthetic constraints) is 1778 [Varkony et al. (1978a)]. Spectroscopic data allow one to look at only the subset of correct molecular weight, e.g., only 264 sterol products have empirical formula $C_{29}H_{48}O$. Additional constraints inferred from spectroscopic data are used by CONGEN during generation. Thus only a small number of CONGEN structures will (typically) remain after one has considered biosynthetic reactions in addition to mass spectral fragmentation patterns and structural constraints inferred from other data.

CHAPTER
FIVE

HEURISTIC DENDRAL PLANNING

Planning before generation is a powerful addition to the generate-and-test method, and is the major design contribution of the DENDRAL system. In the model of scientific discovery embodied by DENDRAL, planning focuses the search on relevant classes of hypotheses and away from irrelevant classes.

5.1 INTRODUCTION

Constraints to keep the generator from producing all possible structures are necessary for all but the simplest problems. We have seen how certain types of constraints can be stated in a manner appropriate for the cyclic generator. In CONGEN these constraints are at present formulated by the user. An early version of the DENDRAL system used a set of planning rules for aliphatic compounds elicited from an expert mass spectrometrist. Each rule specified features of the spectrum that are associated with a particular structure. The planning phase then examined the spectrum for the specific evidence and if it was present the associated structure was added as a GOODLIST constraint. The rules are found in Buchanan, Sutherland, and Feigenbaum (1969). An extension of this planning method, in which rules are automatically generated by the Planning Rule Generator, is described in Buchs et al. (1970a).

The current DENDRAL PLANNER program further automates some aspects of constraint formulation. It embodies a paradigm of mass spectrometry theory that the users instantiate to produce a particular theory for the particular class of compounds with which they are working. This theory is not a sweeping generalization but a collection of specific hypotheses about the likely loci of breaks that will occur when a compound in the class is placed in the mass spectrometer.

Input to the planning phase is: (1) the basic skeleton of the compound class, i.e., the structure common to all members of the class, (2) definitions of the various breaks that might occur, and (3) a mass spectrum (either low or high resolution). The program determines (usually) the molecular ion (using the MOLION program, described below) even if the spectrum does not contain a peak corresponding to it, and formulates constraints that specify which sites on the skeleton are likely, and which unlikely, locations for the nonskeletal groups of atoms. The substructures that are attached to the skeleton are called *substituents*, since they substitute for the hydrogens that would otherwise be there.

5.2 THE EARLY PLANNER AND THE PLANNING RULE GENERATOR

The first class of compounds to which DENDRAL was applied was the amino acids. In predicting the behavior of these compounds in the mass spectrometer, it was assumed that every bond that could break, would break, and they would do so one at a time. This assumption is the *zero-order theory of mass spectrometry*, so-called because it takes into account no site-specific information. This theory proved insufficient for other classes of compounds, although it worked for the amino acids.

More sophisticated versions of planning were used with the acyclic generator applied to the aliphatic ketones. Planning information for these compounds was obtained by careful questioning of mass spectrometrists about ketones in the mass spectrometer, that is, about the kinds of fragmentation processes that were most likely to occur. This knowledge had never been completely codified, however, and thus the effort to specify the information for DENDRAL was a valuable exercise. It proved successful and provided the first version of the system in the full plan-generate-test form where planning was based on planning rules that encompassed more than the trivial zero-order theory. These rules took the form:

Set of *m/e* peaks $-->$ a subgraph

where "$-->$" is read "implies that the molecular structure graph contains."

The successful application to ketones was followed by a similar program for alcohols, with new planning rules for that class of compounds derived in similar fashion. Work was next done for saturated ethers and then saturated amines. At this point similarities were seen among the planning strategies for these classes of compounds. It became possible to develop a general scheme that would generate planning rules for any *saturated acyclic monofunctional (SAM)* compounds, that is, any class with empirical formula of the form $C_n H_{2n+v} x$, where x denotes a heteroatom of valence v. The generalized planning program, based on the Planning Rule Generator, was successfully applied to the saturated acyclic ethers, alcohols, thioethers, thiols, and amines.

The generalized planning program works from a low-resolution mass spectrum and, optionally, a proton NMR spectrum. These are the only inputs necessary; the program is able to determine the empirical formula with a MOLION-like procedure, and thus can determine the class of the compound if it is one of the SAM classes.

The strategy of the Planning Rule Generator is to describe the subgraphs corresponding to all possible configurations of carbons around the heteroatom, and then, assuming only alpha-cleavage fragmentation processes, predict fragmentation products and the combinations of spectral peaks associated with each subgraph. The resulting rules are then used in the same way as manually written rules to determine constraints on generation. This combination, alpha cleavage plus systematic generation of such structures in which alpha cleavage was relevant, proved a successful generalization of the intuitive rules of the human spectrometrists.

As an example, consider the saturated acyclic amines. The heteroatom in this case is N (with a valence of 3) hence there can be one, two, or three carbons adjacent to the heteroatom. Each adjacent carbon can be in one of four forms: it can be of degree 1 (hence part of a methyl radical CH_3), it can be of degree 2 ($—CH_2—$), of degree 3, or of degree 4. Systematic generation of all relevant subgraphs containing the heteroatom and adjacent carbons (some trivial cases, for example, $CH_3—NH_2$, were omitted) yields 31 forms for this class of compounds. Next, each of these 31 cases can undergo an alpha-cleavage fragmentation process in seven ways: one, two, or three alpha cleavages at a time, in all combinations. Each possible process yields a fragment containing the heteroatom, plus one, two, or three fragments containing the remaining atoms of the empirical formula in question. Those fragments that predict peaks actually present in the given mass spectrum then become required (GOODLIST) structures for the generation phase.

These GOODLIST items produced by the planning rules proved able to reduce the set of possible isomers substantially. More remarkable was the additional reduction possible when proton NMR data were also supplied. These data determine the exact number of methyl radicals present in the molecule. This information sometimes was sufficient, in conjunction with the GOODLIST items from the mass spectrometry rules, to constrain the generator to the production of very few candidates. For example, the number of isomers of N,N-dimethyl-1-octadecyl amine is 14,715,813; the number of inferred isomers using the Planning Rule Generator plans and the mass spectrum of that compound is 1,284,792; the number of inferred isomers using both MS and proton NMR planning rules is one![1]

Additional examples are given in Section 8.3.

5.3 MOLION

The first step in planning is to determine the molecular ion. This determination is accomplished by a heuristic program called MOLION [Dromey et al. (1975)], which is also usable as a separate, stand-alone routine. The program works with either a low-resolution or high-resolution spectrum to compute a ranked set of masses (from low-resolution input) or compositions (from high-resolution input) that are candidate molecular ions. It is not always successful, but for 265 compounds covering 10 different classes it has included the correct candidate among its top five choices in 97 percent of the cases; in 89 percent of the cases the correct candidate was among the top three

---

[1] The reason for such dramatic reduction in this case is that the NMR spectrum was interpreted as showing evidence for exactly two methyl radicals, and only a straight chain satisfies this constraint.

choices. The program is limited to compounds of carbon, hydrogen, oxygen, and nitrogen, and depends on a relatively clean and complete spectrum.

If a peak for the molecular ion is present in the spectrum, it will be at highest mass or just below (the peak at highest mass could be the molecular ion with higher mass isotopes of some of its atoms). However, in roughly 20 percent of all spectra there will be no peak in the spectrum corresponding to the molecular ion; in this case the mass of the molecular ion is greater than that of the largest peak present. If we can ignore the possibility of impurities, the peak of next greatest mass after the molecular ion is from a first fragment of the molecule; the difference between it and the molecular ion mass is called a *primary loss*. Other peaks may also be from first fragments of the molecular ion; or they may be from fragments of fragments, called *secondary fragments*. In the latter case, the mass loss suffered by the fragment to produce a secondary fragment is called a *secondary loss*. All interpeak differences are secondary losses except when one of the peaks is the molecular ion peak or when the peaks differ by amounts that could be interpreted as isotopic differences or as transfers of one or more hydrogens. In addition, any low-mass peaks *could* be due to a secondary fragment. Thus the smaller masses are also presumed by the program to be equal in mass to secondary losses.

MOLION is based on the *MOLION postulate: There exists at least one secondary loss in any spectrum that will match a primary loss whether or not the molecular ion is present in the spectrum.* That is, we assume that some composition that breaks from the parent molecule as a primary unit will also break from some fragment as well (see Figure 5-1). MOLION determines all possible secondary losses and adds them to the masses of the larger fragments (those in the upper half of the spectrum) to generate a set of molecular ion candidates. If the MOLION postulate is correct, this set will include the true molecular ion. The final step is to rule out the unlikely candidates and order the remaining ones according to likelihood.

MOLION is organized in a plan-generate-test fashion. The planning stage determines the masses or compositions that are candidate secondary losses. The generate stage combines these in all ways with the larger masses or compositions from the spectrum. The test phase eliminates some of the generated molecular ion candidates and orders the remaining ones by likelihood.

5.3.1 MOLION Planning Phase

This is the most complex part of MOLION. The first step is to simplify the spectrum by converting it to a list of "clusters," sets of peaks separated by less than 3 amu. The three most intense peaks in a cluster are kept and the others are discarded. Then any peak within a cluster that is less than one-third of the largest peak in the cluster is also discarded. This procedure is used to eliminate noise (low-intensity peaks). Simply discarding peaks below a threshold that is the same for the entire spectrum would eliminate some low-intensity but highly informative peaks frequently found at the high-mass end of the spectrum.

The program contains a list of *bad losses*. A bad loss is a mass that can be ruled out on chemical grounds. These normally would include masses 4 to 11, which are too

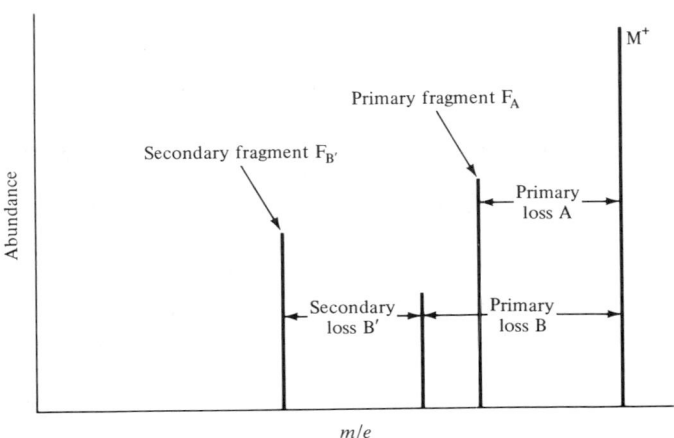

Figure 5-1 The molion postulate. There exists a primary loss A and a secondary loss B' such that A = B', hence M⁺ = F_A + B' (whether or not M⁺ appears in the spectrum).

small to be from atoms found in organic compounds, 12 and 13, which could only be from C and CH, ions that almost never appear, and other higher masses that cannot be formed from elemental masses or that correspond only to compositions that almost never occur. The complete list is

Bad losses. (4 5 6 7 8 9 10 11 12 13 20 21 22 23 24 25 34 37 38 48 49 50 51 53 65 66 76 79 80 81)

Also maintained is a list of *bad compositions* that serves the same function as the bad losses when a high-resolution spectrum is available. The definitions of bad compositions are given in the form of restrictions on the values that may be assumed by the subscripts $x, y, z,$ and n in the formula $C_x H_y O_z N_n$. The restrictions are

Bad compositions:
$$\begin{aligned}
&x < 0 \quad \text{or} \quad y < 0 \quad \text{or} \quad z < 0 \quad \text{or} \quad n < 0 \\
&x > 0 \quad \text{and} \quad y = 0 \quad \text{and} \quad z = 0 \quad \text{and} \quad n = 0 \\
&x = 0 \quad \text{and} \quad y > 2 \quad \text{and} \quad z = 0 \quad \text{and} \quad n = 0 \\
&x = 0 \quad \text{and} \quad y = 0 \quad \text{and} \quad z > 0 \quad \text{and} \quad n = 0 \\
&x = 0 \quad \text{and} \quad y = 0 \quad \text{and} \quad z = 0 \quad \text{and} \quad n > 0 \\
&x = 0 \quad \text{and} \quad n = 0 \quad \text{and} \quad [z > y \quad \text{or} \quad y > 2z] \\
&x \neq 0 \quad \text{and} \quad y > (2x + 3) \\
&x \neq 0 \quad \text{and} \quad z > (x + 1) \\
&x \neq 0 \quad \text{and} \quad y \neq 0 \quad \text{and} \quad x > y + 1
\end{aligned}$$

The program makes first use of the bad losses or bad compositions to determine if there is a molecular ion candidate in the spectrum. The highest-mass peak is compared to its lower neighbors to see if any of the differences correspond to bad losses or bad compositions. A peak that shows no bad losses is a candidate molecular ion. It and other candidates are subjected to further tests later.

The spectrum is then examined to determine the *dominant series*, if one exists. This is the longest series of peaks that differ by multiples of 14 or CH_2. The series will consist either of odd masses or even masses. A dominant odd-mass series implies that the compound contains an even number of nitrogens, perhaps zero, and the mass of the molecular ion must be even. Conversely, a dominant even-mass series means that the compound contains an odd number of nitrogens, and the mass of the molecular ion must be odd. Occasionally there will be even and odd series of equal dominance; in this case no decision is made about nitrogen parity. If the nitrogen parity of the molecular ion mass is determined by this procedure, and the largest mass peak in the spectrum is of opposite parity, we may conclude that no molecular ion peak exists in the spectrum (see Section 2.5.4).

Next, secondary-loss candidates are produced by computing mass differences between all pairs of peaks. Again bad losses and bad compositions are discarded. Other candidates are discarded if they are greater than half the largest mass in the spectrum or otherwise "too large" (an arbitrary limit may be imposed; the default is 115 amu). The remaining set is augmented with all the masses in the lower half of the spectrum, which are presumed to be secondary losses. Finally, for each *even*-mass secondary-loss candidate, the mass of 1 amu less is added to the set of secondary losses because hydrogen transfer is likely to have yielded the (generally stable) even mass whereas the next lower mass was the one actually produced initially.

Each secondary-loss candidate carries a *weighting* to indicate the strength of the evidence supporting its candidacy. This weighting is simply the sum of all the intensities of peaks from which the candidate "arose." For a candidate arising from an inter-peak difference, the weighting is the sum of the intensities of the two peaks; if the same candidate arose from several interpeak differences, all the intensities are summed. Candidates from peaks in the lower half of the spectrum have weighting equal to their intensity. Odd masses arising from even masses one greater adopt their parent's weighting. Those candidates that arise in several ways have the combined weighting from all sources.

The resulting set may then be used as the set of secondary losses, or it may be further reduced by an additional heuristic. This heuristic searches the set of secondary-loss candidates for series that differ by multiples of 14 amu or by one or more CH_2 compositions; only the lower members of each series are chosen as secondary-loss candidates.

The set of secondary losses and the nitrogen parity, if known, are passed to the generating phase.

5.3.2 MOLION Generating Phase

The generating phase is straightforward. Each member of the secondary-loss set is added to each peak in the upper half of the spectrum. Any results having incorrect nitrogen parity are immediately rejected. Any results that are less than the largest peak in the spectrum are also rejected. The sum of the secondary-loss weighting and the upper spectrum peak intensity yields a weighting for each molecular ion candidate. If a candidate arises from more than one combination, its weighting is the sum of its sources' weightings.

5.3.3 MOLION Testing Phase

The basic idea behind the testing phase is to predict for each molecular ion candidate those losses that the candidate must suffer in order to produce the observed spectrum. To the extent the required losses are improbable, the candidate is improbable. The molecular ion candidates are divided by the testing phase into three categories: rejects, unlikelies, and probables.

Differences between each candidate and the prominent peaks in the spectrum are computed. Candidates that must suffer bad losses or bad compositions to produce observed peaks are *rejects*.

Two additional mass lists are kept. The *poor primary-loss* list contains masses that are unlikely to be from first fragments of the molecule. The *poor secondary-loss* list (a subset of the poor primary-loss list) contains masses that are unlikely to be either primary *or* secondary losses. This list is like the bad loss list except that its entries are not chemically impossible, merely unlikely. The lists are

Poor primary losses. (3 14 19 26 27 39 40 54 62 64 67 68 70 82 83 84 86 88 89 90 91 92 93 94 95 96 98 99 103 104 105 106 107 108 109 110)

Poor secondary losses. (3 14 39 64 67 82 84 86 94 95 96 98 103 107 108 109 110)

Candidates that must suffer poor *primary* losses to produce the spectral peak of greatest mass *or* that must suffer poor *secondary* losses to produce other observed peaks are categorized as *unlikelies*.

The remaining candidates are categorized as *probables*. The probable candidates are then ranked by their weightings, normalized to 100. The output is the top members of this list (the number of members to include in the output is user-selected).

5.3.4 An Example of MOLION Performance

We will trace through this procedure for the compound dimethylmalonic acid *n*-butyl ester whose reduced low-resolution spectrum is

Mass	Intensity
41	44
57	48
59	23
70	15
73	12

Mass	Intensity
88	100
115	46
133	14
143	18
144	16
171	95
189	26

5.3.4.1 Planning The dominant series in this spectrum is 59, 73, 115, 143, 171, an odd-mass series. This means the molecular ion has even mass, so it is not present in the spectrum.

We next tabulate all interpeak differences for this spectrum and indicate those that are rejected by virtue of being bad losses or being larger than the default cutoff of 115.

Differences between each peak and lower peaks

Peak	Lower peaks										
	171	144	143	133	115	88	73	70	59	57	41
41											
57											16
59										2	18
70									11*	13*	29
73								3	14	16	32
88						15	18	29	31	47	
115						27	42	45	56	58	74
133					18	45	60	63	74	76*	92
143				10*	28	55	70	73	84	86	102
144			1	11*	29	56	71	74	85	87	103
171		27	28	38*	56	83	98	101	112	114	130!
189	18	45	46	56	74	101	116!	119!	130!	132!	148!

*Indicates bad loss.
!Indicates "too large."

To the set of *all* distinct numbers in this table is added those that are one less than any even member. *Then* the bad losses and redundancies are removed. The union of the resulting set with the set of masses from the lower half of the spectrum, (41 57 59 70 73 88), produces the set of *secondary-loss candidates:* (1 2 3 14 15 16 17 18 27 28 29 31 32 37 41 42 45 46 47 55 56 57 58 59 60 63 69 70 71 73 74 75 83 84 85 86 87 88 91 92 97 98 101 102 103 111 112 113 114). This set has 48 members.

If we were to use only the first few members of each difference-14 series, we would select from this set of series:

Loss series with differences that are multiples of 14

28	42	56	70	84	98	112
15	29	57	71	85		
58	86	114				
18	32	46	60	74	88	102
27	41	55	83			
31	45	59	73	87	101	
47	103					

5.3.4.2 Generating In this example we will use the full set of candidates. Each candidate secondary loss is added to each upper half peak: (115, 133, 143, 144, 171, 189). There are 48 times 6 or 288 such sums, not all distinct. The generator rejects sums less than 189 (the largest peak) and all odd sums, because of the required even nitrogen parity. This reduces the list of molecular ion candidates to 46.

5.3.4.3 Testing From each molecular ion candidate we subtract the largest mass represented in the spectrum, 189, to see what that candidate's primary loss would be. The poor primary-loss list is used to reject bad candidates. Then the next two largest masses represented in the spectrum, 171 and 144, are subtracted from the remaining candidates and any differences that are found in the poor secondary-loss list result in further casualties. These computations are summarized in the following table, which also includes the weighting associated with each candidate.

Candidate M$^+$	Less 189	Less 171	Less 144	Weighting
190	1	19	46	392
192	3**	21*	48	49
196	7*	25*	52*	29
198	9*	27	54	1154
200	11*	29	56	857
202	13*	31	58	556
204	15	33	60	318
206	17	35	62	725
208	19**	37	64***	171
212	23*	41	68	254
214	25*	43	70	73
216	27**	45	72	1504
218	29	47	74	845
220	31	49	76*	368
224	35	53*	80	58
226	37	55	82***	947
228	39**	57	84***	431
230	41	59	86***	488
232	43	61	88	100
234	45	63	90	626
236	47	65*	92	888
240	51*	69	96***	152
242	53*	71	98***	135
244	55	73	100	1175
246	57	75	102	469

Candidate M⁺	Less 189	Less 171	Less 144	Weighting
248	59	77	104	49
252	63	81	108***	29
254	65*	83	110***	359
256	67**	85	112	366
258	69	87	114	352
260	71	89	116	28
262	73	91	118	333
264	75	93	120	62
268	79*	97	124	107
272	83**	101	128	534
274	85	103***	130	165
276	87	105	132	164
280	91**	109***	136	58
282	93**	111	138	118
284	95**	113	140	143
286	97	115	142	107
290	101	119	146	298
292	103**	121	148	60
300	111	129	156	118
302	113	131	158	143

*Denotes bad loss.
**Poor primary loss.
***Poor secondary loss.

Of the 18 acceptable candidates, 244 has the largest weighting and is in fact known to be the mass of the molecular ion of this compound.

5.3.5 Optional Heuristics in MOLION

If the program as just described fails to narrow the list of candidates sufficiently, there are additional options.

1. Candidates may be rejected if there is no metastable ion support, that is if they show no true daughter ion peaks.
2. A spectrum obtained using a low ionization voltage normally will show a strong relative peak at the molecular ion because other fragmentations are less likely. This will be the case even though a spectrum obtained at normal voltages does not show this peak because the greater intensities of other peaks place it in the noise level. Candidates that show *no* peak at low ionization voltage may be rejected.
3. If there is a pair of candidates, differing by one or two H's, the lower may be automatically rejected.
4. If there is a pair of candidates, the higher of which is less intense and can be accounted for by variations in abundance of natural isotopes, that higher one may be automatically rejected.
5. The user may insist that the molecular ion is present in the spectrum if that is generally true for compounds of the class in question. (This is true of the estrogens, for example.)

5.4 EMPIRICAL FORMULA

If MOLION has selected a molecular ion candidate from high-resolution data, then the composition, that is the empirical formula, is known. If only a low-resolution spectrum was available, further work must be done to determine the empirical formula from the mass of the molecular ion. At present DENDRAL does not use the program that was written to perform this determination; the empirical formula is determined by other means (often high-resolution mass spectrometry) and supplied to the PLANNER by the user. The reason is that the program can never produce a unique answer when both nitrogen and oxygen are possible, since $-NH_2$ and $=O$ both have nominal mass contributions of 16 amu.

5.5 GENERALIZED BREAK ANALYSIS

For many analytical problems, it is known what class of compounds is under investigation, particularly when the source of the material is known and the empirical formula of the molecular ion has been determined. If chemists have experience with this class, they may be able to predict the fragmentations, or breaks, that are most likely to have occurred. This class-specific information can be used to produce constraints that will reduce the DENDRAL generator's output.

Breaks are defined by specifying (1) the bonds that break, (2) the charge location, (3) any accompanying hydrogen transfers, and (4) any accompanying losses of small molecules such as water. Breaks are represented schematically with wavy lines as illustrated in Figure 5-2.

From the empirical formula and the definition of the class skeleton, the program can readily determine the numbers and types of atoms and unsaturations that are not part of the skeleton and which thus are available to form the substituent structures that distinguish the particular compound under investigation. This capability is shown in a simplified example in Figure 5-3. It remains to determine the exact composition of each substituent and its location on the skeleton. To do this determination, the mass spectrum is consulted for evidence that can be associated with each of the hypothesized breaks. In general, all these breaks may have occurred, but the fragments resulting from a particular break will have compositions that differ depending on how the substituent atoms are divided among the fragments. The program determines for each break all the ways the substituent atoms may be divided and the evidence in the spectrum associated with each division. This process of course does not determine the

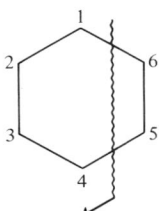

Figure 5-2 Schematic representation of a fragmentation. This fragmentation involves breaking bonds between atoms 1 and 6 and between atoms 4 and 5. The arrow indicates (positive) charge placement on the larger fragment. No hydrogen transfers or neutral molecule losses are indicated in this example.

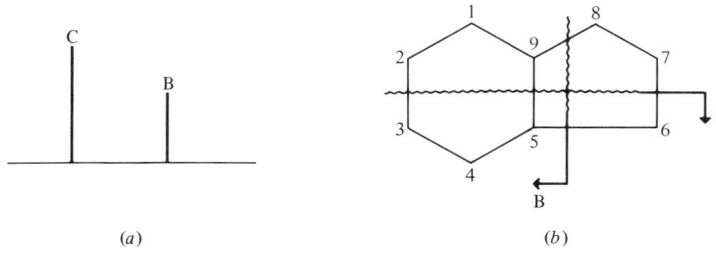

Figure 5-3 Simplified example of reasoning by the DENDRAL planner. (*a*) The mass spectrum contains a peak corresponding to the mass-to-charge ratio of fragment C *without* an OH substituent and a peak corresponding to the mass-to-charge ratio of fragment B *with* an OH substituent. (*b*) Thus the DENDRAL planner is able to infer that the OH substituent is probably in the part of fragment B that is distinct from fragment C, i.e., at node 1, 2, or 9.

exact location or composition of a substituent. However, combining the information from all the breaks further narrows the possible arrangements.

The DENDRAL PLANNER has been successfully applied to the estrogens, and this application will serve as an illustration of its potential. The user first defines the characteristic skeleton of the class using the structure-editing commands of CONGEN. As illustrated in the example at the end of the last chapter, these commands permit the definition of rings and chains of carbon atoms; these may be modified by renaming some atoms (to other atom types), by increasing the bond order between atoms (with JOIN a b), by growing chains from given atoms (BRANCH a n creates a chain of *n* atoms from atom *a*), and by setting the free valences of atoms. The structure being formed may be drawn (on the terminal) with either letter names for atoms (ADRAW) or numbers for each atomic position (NDRAW). The following is a record of a session in which the estrogen skeleton is constructed (and named EST). User-typed commands follow the system prompt symbols "#" and ">", or are replies to questions. They are in bold type. Other typing was done by the program.

WELCOME TO THE DENDRAL PLANNER.
TYPE ? AT ANY TIME (REALLY!) FOR HELP IN INTERPRETING WHAT THE PROGRAM WANTS.
USE <CNTRL>-E (PRESS THE CNTRL AND E KEYS SIMULTANEOUSLY) TO GET TO COMMAND MODE. THIS USUALLY CANCELS THE EFFECT OF THE PREVIOUS COMMAND.
CLASS NAME: **EST**
GET CLASS PARAMETERS FROM A FILE (Y/N) ? **N**
READ A NEW MASS SPECTRUM (Y/N) ? **N**

START PLANNING (Y/N)? **N**
WANT HELP (Y/N)? **N**
EDIT STRUC
(NEW STRUCTURE)
> **CHAIN 10**

>JOIN 1 2 3 4 5 10 5 10 10 1

[Comment: The partially built structure is drawn to see where to add the next piece.]

>NDRAW

```
EST
   6    4
  / \  / =
 7   5   3
 |   =   |
 8  10   2
  \ /  \ =
   9    1
```
>BRANCH 9 7

[Comment: Again, the partial structure is drawn before adding the final specifications.]

>NDRAW

```
EST
   3-2
   = =
  4   1
   \ /
   5=10
   / \
  6   9-1-1-1-1-1-1-17
   \ /  1 2 3 4 5 6
   7-8
```
>JOIN 14 8 17 13
>BRANCH 13 1
>NDRAW

```
EST
             6    4
            / \  / =
           7   5   3
           |   =   |
       15  8  10   2
       / \ / \ / \ =
      16  14  9   1
       |   |  |
      17---13 11
            / \ /
           18  12
```
>DONE

SKELETAL COMPOSITION: ((H . 24) (C . 18))

Next we enter our "theory" of the mass spectrometry of estrogens. This entry takes the form of specifying possible breaks (done by listing the names of bonds, i.e., node pairs, that are broken), by indicating which fragment will bear the charge (done by naming any atom in the ion), and indicating how many hydrogen atoms might migrate upon fragmentation (each potential hydrogen transfer is indicated by a signed number denoting the number of possible transfers *to* the ion; there may be more than one possibility).

Note that the form of the planning rules has changed from the early version of the planning program. Instead of requiring sets of diagnostic peaks associated with every possible configuration of relevant subgraphs, the program now accepts rules of the form: subgraph ==> break, interpreted as "when the graph contains the given subgraph, infer that the particular break will occur." These rules are more natural to the chemist and constitute the codification of fragmentation knowledge commonly used in texts. With this change, the Planning Rule Generator was subsumed in the planning procedure itself.

For the estrogen example we have the break definitions in Table 5-1, pictured in Figure 5-4.

This information is given to the PLANNER in the following dialogue. The program's one-word requests for information terminate with ":" and the user's replies follow immediately in bold type. The user has typed "?" at certain places to elicit explanations from the program to make the dialogue self-explanatory.

Table 5-1 Estrogen breaks

Name	Break definition	Transfer hydrogens	Charge on:
B	(14.15)(13.17)	(−1, 0)	14
C	(9.11)(14.13)(15.16)	(−1, 0)	9
D	(9.11)(14.13)(16.17)	(−2, −1)	9
E	(11.12)(8.14)	(−1, 0)	11
F	(9.11)(8.14)	(−1, 0)	9

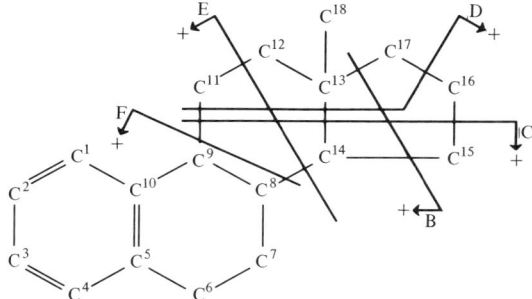

Figure 5-4 The estrogen skeleton with break definitions.

BREAKS
LABEL: B
BOND: ?

ENTER THE NUMBERS OF THE ATOMS BETWEEN WHICH THE BOND IS BROKEN.
FOR EXAMPLE,
BOND: 3 7
IF THE BREAK IS BETWEEN ATOM 3 AND ATOM 7. YOU MAY ENTER ONE PAIR OF NUMBERS PER LINE.
THE FIRST ATOM MENTIONED (IN THIS CASE 3) SHOULD BELONG TO THE CHARGED FRAGMENT.
SIMPLY USE CARRIAGE RETURN WHEN YOU ARE FINISHED ENTERING BONDS FOR THIS BREAK.
BOND: 14 15
BOND: 13 17
BOND:
HTRANS: ?

ENTER A LIST OF NUMBERS WHICH CORRESPONDS TO THE NUMBER OF HYDROGENS TRANSFERRED INTO OR OUT OF THE CHARGED FRAGMENT. (NEGATIVE IF HYDROGENS MIGRATE OUT.) FOR EXAMPLE -
HTRANS: 1 0 -1
INDICATES THAT ONE HYDROGEN IS TRANSFERRED OUT, OR ONE IS TRANSFERRED INTO THE CHARGED FRAGMENT, OR NONE MOVE AT ALL.
HTRANS: -1 0
LABEL: C
BOND: 9 11
BOND: 14 13
BOND: 15 16
BOND:
HTRANS: -1 0
LABEL: D
BOND: 9 11
BOND: 14 13
BOND: 16 17
BOND:
HTRANS: -2 -1
LABEL: E
BOND: 11 12
BOND: 8 14
BOND:
HTRANS: -1 0

LABEL: **F**
BOND: **9 11**
BOND: **8 14**
BOND:
HTRANS: **-1 0**
LABEL:

[Comment: The user now asks to see the value of BREAKS as a check.]

BREAKS?
....
LABEL: B
BONDS: ((14 . 15) (13 . 17))
HTRANS: (-1 0)
....
LABEL: C
BONDS: ((9 . 11) (14 . 13) (15 . 16))
HTRANS: (-1 0)
....
LABEL: D
BONDS: ((9 . 11) (14 . 13) (16 . 17))
HTRANS: (-2 -1)
....
LABEL: E
BONDS: ((11 . 12) (8 . 14))
HTRANS: (-1 0)
....
LABEL: F
BONDS: ((9 . 11) (8 . 14))
HTRANS: (-1 0)
....

SAVE
FILE NAME: **PLANNER.EST**
PLANNER.EST;1 HAS BEEN SAVED.
DONE
BYE

5.5.1 Producing Constraints

The first step is to predict for each break what possible ions could result. In general, ions possible for a given break have compositions consisting of at least the skeletal fragment (less any net losses of hydrogen) and at most the skeletal fragment plus all substituents (plus any possible net gains of hydrogen). Computing possible ions for each defined break is thus straightforward. In our example, consider estrogen break B and a compound with empirical formula $C_{18}H_{24}O_2$. Break B will leave an ion containing 15 carbons and 18 hydrogens. Or it may contain 15 carbons and 17 hydrogens, since the break definition permits one migration of H away from the ion. The substituent atoms are two oxygens. Either, both, or neither of these could be on the ion; in each of these three cases 17 or 18 hydrogens could remain. Hence six possible ions could result from break B. They are organized into three ion series, where ions are in the same series if they differ in composition only by possible hydrogen transfers:

$$C_{15}H_{18}, \quad C_{15}H_{17}$$
$$C_{15}H_{18}O, \quad C_{15}H_{17}O$$
$$C_{15}H_{18}O_2, \quad C_{15}H_{17}O_2$$

The next step is to find evidence in the spectrum for each of the possible ions and sum the evidence within each ion series. The user has other optional constraints which may be applied. For example, the intensities of the corresponding peaks may be required to exceed a threshold; the threshold may be set differently for each break if the chemist believes that certain breaks always give prominent peaks.

After this analysis we have for each break an ion series that has greatest empirical support. The next step is to consider each possible combination[2] of these series with a Chinese menu algorithm: one from break B, one from break C, and so forth. Some of these combinations may be impossible, and others may restrict the substituents to a single locus or a range of them.

For example, a spectrum actually produced from estradiol, the isomer of $C_{18}H_{24}O_2$ in which the oxygens are at nodes 3 and 17, should contain strongest evidence for the single-oxygen series for each of the five break definitions. Accordingly, PLANNER can determine the following. Exactly one oxygen is on one of the nodes 1 to 10 (from break F evidence); nodes 11, 14, 15, and 16 do not have an oxygen (else break D or E would have yielded double-oxygen series evidence); nodes 12, 13, and 18 do not have an oxygen (else break B would have yielded double-oxygen series evidence). Thus the second oxygen must be at node 17. These constraints will substantially reduce the number of isomers of $C_{18}H_{24}O_2$ that the generator would otherwise produce.

The constraints thus determined are suitable for use by the GOODLIST and BADLIST mechanisms of the generator. Additional information may also be supplied with GOODLIST and BADLIST. For example, naturally occurring estrogens typically have an oxygen at a particular site on the skeleton (node 3); if the compound is known to occur naturally, this structural feature may be added to GOODLIST. In addition, the program makes a thorough analysis of any proposed structures and rules out all those that would violate valence constraints.

If no possible structures remain, that is, if the constraints rule out all possibilities, the PLANNER relaxes its criteria on the empirical evidence and tries again.[3]

---

[2]This combinatorial generator under constraints bears surface similarities to CONGEN but is less general.

[3]This capability, too, is under the control of the user, giving a significant degree of control over the feedback criteria. For example, the user may wish breaks C and D to be relaxed first since they are less reliable than the other breaks.

5.6 CONCLUSION

The DENDRAL PLANNER can be useful as a stand-alone program, for it may reduce the number of possibilities to the point where there is no need to generate them by computer. More generally, however, the constraints produced by PLANNER, along with any others that the user wishes to apply for whatever reason (and from whatever source), are passed to the generator.

It is clear from this description that PLANNER will be of assistance only when we are explaining a mass spectrum from a class of compounds for which we have some prior experience that permits the definition of reasonable mass spectrometric breaks. This is frequently not the case. Other types of planning—constraint selection—may then be needed. Often the only solution is to specify them "by hand" directly to CONGEN.

The PLANNER, and its component MOLION, are good illustrations of the design of the DENDRAL system and the general philosophy of design that characterizes the project. No explicit formal theory underlies these programs. Describing them, even in the idealized fashion we have employed in this chapter, is at best clumsy. However, their ad hoc character is their strength when combined with program flexibility. Since so much of any scientist's knowledge is intuitive and judgmental, it cannot be readily formalized. Programs such as those we have just described nonetheless permit this knowledge to be employed in useful ways. The processing speed and clerical superiority of the computer in fact amplify, in many cases, the power of the scientist's informal knowledge. A particular case in point is the PLANNER's analysis of spectra taken from unseparated mixtures of estrogens, described in Smith et al. (1973a) and in Chapter 8. The record-keeping and cross-checking demands of this task are too great to permit a thorough job by hand.

CHAPTER
SIX

HEURISTIC DENDRAL TESTING

Verification is the best understood part of the scientific enterprise, but it is usually discussed outside the context of discovery. In DENDRAL, the verification process is an important stage in the discovery process, for by testing the generated hypotheses the program can better select the most plausible ones. DENDRAL's testing program contains a predictive theory of mass spectrometry from which testable predictions are made. Some hypotheses can be discarded if their predicted data are inconsistent with observation; the rest are ranked in order of plausibility.

6.1 INTRODUCTION

If planning were perfect and generating were exhaustive, no spurious solution candidates would be produced and no solutions would be missed. DENDRAL planning, however, is not perfect and so a testing phase is applied to each candidate.

There are two reasons why planning cannot be perfect even in principle. First there may be tests that can only be applied to fully specified solution candidates or to *sets* of them. Second there may be tests that are too costly to apply in the planning phase but are not too expensive to apply to the smaller set of solution candidates. Both these factors operate with DENDRAL.

The constraints that DENDRAL planning produces are conditions that exclude or specify classes of structures defined by the presence of certain substructure features or global properties. A molecular structure that meets all the constraints may nonetheless have properties by which it can be recognized as a spurious solution. Furthermore there are ways in which candidates can be ranked, and rankings necessarily involve comparisons among sets of candidates and can thus not be formulated as a priori constraints on generation.

Some of the tests that can be used to reject proposed structures require data from metastable defocusing experiments (see Section 2.5.2). These data are expensive to collect. Therefore the experiments are done only for those cases that pass the plan-generate phases of DENDRAL.

The program that tests the generator's output might have been called the TESTER because of the way it is used, but it has been called the PREDICTOR because of what it does. This program is driven by a set of rules, called *productions*, that define a theory of the behavior of compounds in a mass spectrometer. The productions state that structures of a given form will fragment in certain ways. They are thus like the rules used in PLANNER, except that the format is more general. Given a set of productions, PREDICTOR applies them to a proposed structure graph to predict what ions will be produced; the productions are applied recursively to all ion graphs they produce. The result is a predicted mass spectrum that can be compared to the actual data from which DENDRAL began. Structures that produce spectra in close agreement with the data are ranked highly. PREDICTOR also predicts metastable peaks, and further laboratory work may be able to distinguish among the best final candidates.

6.2 PREDICTOR PRODUCTION SYSTEM

PREDICTOR is based on a *production system* control structure [Newell (1972)]. A *production* is a rule that defines a *situation* and an *action* to be taken when that situation exists. The control structure is a regimen for the application of productions and for recording and processing the results of the application.

The situation-part of a PREDICTOR production describes a chemical graph. The action-part of the production consists of a set of operations that alter the graph to produce other graphs. The action-part may contain a complete production as a component, so that some operations are performed conditionally. The operations that PREDICTOR permits are any INTERLISP functions (Section 3.4); in fact they define fragmentations, hydrogen transfers, and means for computing relative abundances of the products of the fragmentation processes.

The PREDICTOR control structure applies the production rules to a set of ions. First a molecular ion is constructed and placed as the only item on an *ion list*. The set of productions is scanned, and each one that is applicable is applied to produce one or more new ions that are then added to the end of the ion list. When all applicable productions have been applied, the current ion is put into the *spectrum list* and the next ion on the ion list is selected. When all ions on the ion list have finally been processed (or the maximum permitted depth of recursion has been reached), the procedure is complete. The spectrum list then corresponds to a mass spectrum, since each entry is an ion that has a mass or composition, and each has associated with it a number that denotes its abundance. Note that some ions may have several entries on the spectrum list if they arose from more than one fragmentation process. The abundances are accumulated and normalized to produce the mass spectrum prediction. For each peak other than the molecular ion a metastable peak (M*PEAK) is computed. [Recall that if d is the mass of the (daughter) peak and p is the mass of the (parent) molecular ion, then the corresponding metastable peak is at mass d^2/p. See Section 2.5.2.]

6.3 GRAPH STRUCTURE AND PRODUCTION REPRESENTATION

While it is not necessary to know the details of the INTERLISP data structures used by the program to understand its logic, it will be convenient to introduce some of these details because they enable descriptions to be given in linear form that is both more compact and more convenient to reproduce than graphs.

6.3.1 Representation of Chemical Graphs

Each node of a chemical graph is given a number and a name. These are redundant and both exist only for reasons having to do with INTERLISP conventions, but we include both in our examples so that they correspond to program listings. The node is described by indicating the type of atom, its connectivity to other nodes, the number of unsaturations (double bonds) from that node (these are called "dots"), and the number of hydrogens attached to the node. Connections are indicated by a list of node names enclosed in parentheses. The entire node description is enclosed in parentheses. Finally the list of nodes defining a structure is enclosed in parentheses along with the name of the structure.[1]

To illustrate, consider C_6H_5OH. We arbitrarily use the number of a node as its name.

NODE DIAGRAMS

```
                                    H   H
            6 = 7                   C = C
           /     \                 /     \
          5       1 - 2          HC       C - OH
          =       =               =       =
          4   -   3               C   -   C
                                  H       H
```

ATOM NAME	ATOM TYPE	NODE NUMBER	NEIGHBORS	DOTS	NUMBER OF HYDROGENS
1	C	1	(2 3 7)	1	0
2	O	2	(1)	0	1
3	C	3	(1 4)	1	1
4	C	4	(3 5)	1	1
5	C	5	(4 6)	1	1
6	C	6	(5 7)	1	1
7	C	7	(1 6)	1	1

[1] The original representation of a chemical graph was as a LISP list with sublists indicating branching. A second notation was as a list of LISP atoms [for example, (C1, C2, C3)], each with its own property list indicating the properties of NEIGHBORS, NUMBER OF HYDROGENS, DOTS, etc. The representation described here is faster and more compact because it uses fixed positions in lists instead of explicit names and values.

The complete notation is (spacing on the page is irrelevant and is chosen for readability):

(PHENOL
        (1 C 1 (2 3 7) 1 0)
        (2 O 2 (1) 0 1)
        (3 C 3 (1 4) 1 1)
        (4 C 4 (3 5) 1 1)
        (5 C 5 (4 6) 1 1)
        (6 C 6 (5 7) 1 1)
        (7 C 7 (1 6) 1 1)
)

The same notation is used to describe classes of structures in which some of the information is not completely specified. In particular, substructures are defined in such a way that their connections to larger structures are not uniquely defined. The symbol X when used in a connectivity list denotes exactly one connection to a node *outside* the subgraph. The symbol "--" means "don't care." When used in the position of a node, "--" denotes zero or more connections outside the substructure; when used instead of an integer as a dot or hydrogen-count parameter, it indicates any integer value, including zero. When actual numbers are not needed by INTERLISP, X may also appear as a node number, as is the case with substructures. Thus the ketone substructure is

(KETONE
        (1 C X (2 X X) 1 0)   X-C-X
        (2 O X (1) 1 0)   =
)   O

The alcohol substructure is

(ALCOHOL
        (1 C X (2 X --) --)   X-C-OH
        (2 O X (1) 0 1)
)

6.3.2 Representation of Productions

Each production consists of a situation-part and an action-part. The situation-part is a predicate, which is either true or false of the structure under consideration. For example, the predicate (WHERE X) evaluates to true if X is a substructure of the *active ion*, that is, the one to which the production is being applied. The special argument-free predicates NIL and DEFAULT may be used as the situation-part. NIL always evaluates to false; this predicate is used to take a production out of action temporarily without actually deleting it from the production set. DEFAULT evaluates to true if and only if all previously tried productions have proved inapplicable.

Other situation-part predicates are specific to chemistry. DB(N,M) is true if there is a double bond between nodes numbered N and M. ISIT(S) is true if substructure S is contained in the active ion. MISSING(N,M) is true if nodes N and M are not connected. ON(S, N) is true if substructure S is contained in the active ion, with node N substituted for any unspecified atoms in substructure S. ISIT, ON, and WHERE, when true, produce as side effects an indication of the location of the match. This information is available for use by functions in the action-part of the production.

The action-part of a production is more complex. It may have more than one component. Each component may itself be a production. A component also may be the name of a function that will be executed, selecting its parameters from the current context. Finally, a component may be a list of two or three functions specific to mass spectrometry. The *break function* produces new ions from the active ion and adds them to the ion list. The *intensity function* determines an intensity for each newly formed ion. It may be a number indicating the relative abundance of the parent ion to be assigned to the daughter ion. The *transfer function* produces additional ions according to rules of hydrogen transfer but does not add these to the ion list. (The transfer function is optional and may be absent.) Each function-list component has a label that is used in the output to indicate the source of each spectral peak. The label is the first symbol in the function list.

To summarize, a PRODUCTION is (SITUATION-PART ACTION-PART). A SITUATION-PART is one of the following three forms:

1. (ANY LISP PREDICATE, for example, ISIT, WHERE, DB, ON)
2. (NIL)
3. (DEFAULT)

An ACTION-PART is a list of ACTION-PART COMPONENTS. An ACTION-PART COMPONENT is one of the following three forms:

1. (LABEL BREAK-FUNCTION INTENSITY-FUNCTION TRANSFER-FUNCTION (optional))
2. ANY LISP FUNCTION
3. A PRODUCTION

The following production for estrogens is a production rule formulation of the theory of estrogen fragmentation that was presented in the discussion of PLANNER (see Section 5.5). The break labels, node numbers, and hydrogen-transfer information are the same as used in that discussion. The *intensity-function* of each component of the production is here the integer 100, which simply means that any generated ion will be assigned 100 percent of the intensity of the ion from which it derives. The estrogen production has five function-list components, each function list associated with a particular break definition; the break name is used as the label of its function list. The functions BREAKBND and HTRANS are a break function and a transfer function, respectively, to break a bond, or a set of bonds, and to transfer hydrogens.

((WHERE ESTROGEN)
  (B (BREAKBND((14 . 15) (13 . 17))   100 (HTRANS -1 0))
  (D (BREAKBND((9 . 11) (14 . 13) (16 . 17)))   100 (HTRANS -2 -1))
  (C (BREAKBND((9 . 11) (14 . 13) (15 . 16)))   100 (HTRANS -1 0))
  (E (BREAKBND((11 . 12) (8 . 14)))   100 (HTRANS -1 0))
  (F (BREAKBND((9 . 11) (8 . 14)))   100 (HTRANS -1 0))
)

A syntactically more complex production is one that defines McLafferty rearrangement, defined in an earlier chapter (Section 2.5.3). The structures that undergo this rearrangement contain the substructure named GRAFMC, shown below (compare Figure 2-13).

(GRAFMC
       (1 C X (2 4 --) 1)
       (2 -- X (1 --) 1)
       (4 -- X (1 5 --))
       (5 -- X (4 6 --))
       (6 C X (5 --))
)

The production is labeled MCLAFFERTY. It corresponds to an MS process that occurs in many cases and has thus been defined and named so that it may be referenced in defining class-specific productions. A number of such cases are gathered together below under the heading SUB-PRODUCTIONS. These processes are used in the subsequent definitions of the PREDICTOR productions. The situation-part of the MCLAFFERTY production is (WHERE GRAFMC). The action-part of MCLAFFERTY uses six functions. Two of them, LASTION and LASTINT, simply return the previous ion and its intensity, respectively. Three of them, MRRFGT, MRRFGT2, and MRRFGT3, produce the required hydrogen migrations before performing breaks that produce new ions. The final function GAMMACLEAVAGES, produces (possibly) several ions by producing all gamma cleavages (see Section 2.4.1).

The other subgraphs referenced in the MCLAFFERTY production are defined below. The integers used as intensity-functions are interpreted as before: the intensity of the generated ion is the indicated percentage of the intensity of the current ion. The MCLAFFERTY production is

(MCLAFFERTY
  ((WHERE GRAFMC) ((WHERE GRAFMCMETHYL)
               (MCLAFFCH3 MRRFGT3 100)
               (MCLAFFERTY MRRFGT 200))
      ((WHERE GRAFDBLMCALPHA)
          (DBLMCLAFF MRRFGT3 100)
          (DBLMCLAFF MRRFGT2 100)
          (MCLAFFERTY MRRFGT 200))

                (DEFAULT
                        (MCLAFFERTY MRRFGT 200))
                ((WHERE GRAFMC5)
                        (MCLAFFERTY+1 LASTION   (LASTINT 10)   (HTRANS 1))
                        (GAMMA GAMMACLEAVAGES (LASTINT 100)))
))

The following productions comprise a partial theory of mass spectrometry as embodied in one version of the PREDICTOR. The functions they employ are defined subsequently. The chemical graphs they reference are shown in Table 6-1. Adding new productions is simple as long as no new functions are introduced.

(* PRODUCTIONS
------------------------)
SITUATION–PART   ACTION–PART

    ((WHERE KETONE) MCLAFFERTY
            (MAJORALPHA MAJORALPHACLEAVAGE MAJORALPHAINT)
            (MINORALPHA MINORALPHACLEAVAGES MINORALPHAINT)
             COELIM
)
    ((WHERE ALDEHYDE) MCLAFFERTY2
            (MAJORALPHA MAJORALPHACLEAVAGE MAJORALPHAINT)
)
    ((WHERE ALCOHOL) H2OELIM
            (MAJORALPHA MAJORALPHACLEAVAGE MAJORALPHAINT)
            (MINORALPHA MINORALPHACLEAVAGES MINORALPHAINT)
             (* DELETEPARENT 0)
)
    ((WHERE THIOL) H2OELIM
            (MAJORALPHA MAJORALPHACLEAVAGE MAJORALPHAINT)
            (MINORALPHA MINORALPHACLEAVAGES MINORALPHAINT)
             (* DELETEPARENT 0)
)
    ((WHERE ETHER) (MAJORALPHA MAJORALPHACLEAVAGE
                        MAJORALPHAINT)
            (MINORALPHA MINORALPHACLEAVAGES MINORALPHAINT)
            ((WHERE GRAFCH3ALPHA) (ALPHA-H REMOVEH1
                 (LASTINT 100)))
            ((GREATERP LEVEL 0)
                 ((WHERE GRAFETH1) (REARR AMINERR 200)))
)
    ((WHERE THIOETHER) (MAJORALPHA MAJORALPHACLEAVAGE
                        MAJORALPHAINT)

```
              (MINORALPHA MINORALPHACLEAVAGES MINORALPHAINT)
              ((WHERE GRAFCH3ALPHA) (ALPHA-H REMOVEH1
                     (LASTINT 100)))
              ((GREATERP LEVEL 0)
                     ((WHERE GRAFETH1) (REARR AMINERR 200)) )
     )
     ((WHERE AMINE) (MAJORALPHA MAJORALPHACLEAVAGE
                            MAJORALPHAINT)
              (MINORALPHA MINORALPHACLEAVAGES MINORALPHAINT)
              ((WHERE GRAFCH3ALPHA) (ALPHA-H REMOVEH1
                     (LASTINT 100)))
              ((GREATERP LEVEL 0)
                     ((WHERE GRAFETH1) (REARR AMINERR 200)) )
     )
     ((WHERE OXIME) (MAJORALPHA MAJORALPHACLEAVAGE
                            MAJORALPHAINT)
              (GAMMA GAMMACLEAVAGES 200)
              MCLAFFERTY
              H2OELIM
              COELIM
     )
     ((WHERE ESTROGEN)
         (B (BREAKBND((14 . 15) (13 . 17))   100 (HTRANS -1 0))
         (D (BREAKBND((9 . 11) (14 . 13) (16 . 17)))   100 (HTRANS -2 -1))
         (C (BREAKBND((9 . 11) (14 . 13) (15 . 16)))   100 (HTRANS -1 0))
         (E (BREAKBND((11 . 12) (8 . 14)))   100 (HTRANS -1 0))
         (F (BREAKBND((9 . 11) (8 . 14)))   100 (HTRANS -1 0))
     )
(*  SUB-PRODUCTIONS
------------------------)
(CH4ELIM
     ((WHERE GRAFCH41*3) (CH4ELIM1*3 ELIM 40))
     ((WHERE GRAFCH41*4) (CH4ELIM1*4 ELIM 50))
)
(COELIM
     ((WHERE GRAFCO) (COELIM LOSE1*2 50))
)
(H2OELIM
     ((WHERE GRAFH2O1*2) (H2OELIM1*2 ELIM 25))
     ((WHERE GRAFH2O1*3) (H2OELIM1*3 ELIM 80))
     ((WHERE GRAFH2O1*4) (H2OELIM1*4 ELIM 100))
)
(H2SELIM
     ((WHERE GRAFH2S1*3) (H2SELIM1*3) ELIM 8))
     ((WHERE GRAFH2S1*4) (H2SELIM1*4 ELIM 10))
```

)
(HALOGENELIM
    ((WHERE GRAFHAL1*3) (HALOGENELIM1*3 ELIM 40))
    ((WHERE GRAFHAL1*4) (HALOGENELIM1*4 ELIM 50))
)
(NH3ELIM
    ((WHERE GRAFNH31*3) (NH3ELIM1*3 ELIM 40))
    ((WHERE GRAFNH31*4) (NH3ELIM1*4 ELIM 50))
)
(MCLAFFERTY
    ((WHERE GRAFMC ((WHERE GRAFMCMETHYL)
                (MCLAFFCH3 MRRFGT3 100)
                (MCLAFFERTY MRRFGT 200))
        ((WHERE GRAFDBLMCALPHA)
                (DBLMCLAFF MRRFGT3 100)
                (DBLMCLAFF MRRFGT2 100)
                (MCLAFFERTY MRRFGT 200))
        (DEFAULT
                (MCLAFFERTY MRRFGT 200))
        ((WHERE GRAFMC5)
                (MCLAFFERTY+1 LASTION   (LASTINT 10)
                (HTRANS 1))
                (GAMMA GAMMACLEAVAGES (LASTINT 100)))
))

The functions used by these productions have the following effects. Their definitions, of course, are in INTERLISP.[2]

BREAK FUNCTIONS

AMINERR

1. Find the bonds alpha to node 2.
2. Break one major bond from step 1.
3. Break one major bond adjacent to node 2.
4. Place a hydrogen at node 2.
5. Return a list containing the ion containing node 2.

BRK (BONDSET)

1. Break each of the bonds in bondset.
2. Return a list containing the ion containing the first node of the first bond in bondset.

[2] This is an example of *procedural imbedding of knowledge*, as it is now called in the artificial intelligence literature.

Table 6-1 Structure graphs referred to by the example productions

```
(*STRUCTURE  GRAPHS
--------------------)
(KETONE
         (1 C X (2 X X) 1 0)
         (2 O X (1) 1 0)
)
(ALDEHYDE
         (1 C X (2 X) 1 1)
         (2 O X (1) 1 0)
)
(ALCOHOL
         (1 C X (2 --) --)
         (2 O X (1) 0 1)
)
(THIOL
         (1 C X (2 --) --)
         (2 S X (1) 0 1)
)
(ETHER
         (2 O X (X X) 0 0)
)
(THIOETHER
         (2 S X (X X) 0 0)
)
(AMINE
         (2 N X (X --))
)
(OXIME
         (1 C X (2 --) 1)
         (2 N X (1 3) 1 0)
         (3 O X (2 X) 0 0)
)
(ESTROGEN
         (1 C X (2 10 --) 1)
         (2 C X (1 3 --) 1)
         (3 C X (2 4 --) 1)
         (4 C X (3 5 --) 1)
         (5 C X (4 6 10) 1 0)
         (6 C X (5 7 --) --)
         (7 C X (6 8 --) --)
         (8 C X (7 9 14 --) --)
         (9 C X (8 10 11 --) --)
         (10 C X (1 5 9) 1 0)
         (11 C X (9 12 --) --)
         (12 C X (11 13 --) --)
         (13 C X (12 14 17 18) 0 0)
         (14 C X (8 13 15 --) --)
         (15 C X (14 16 --) --)
         (16 C X (15 17 --) --)
         (17 C X (13 16 --) --)
         (18 C X (13 --) --)
)
(GRAFMC
         (1 C X (2 4 --) 1)
         (2 -- X (1 --) 1)
         (4 -- X (1 5 --))
         (5 -- X (4 6 --))
         (6 C X (5 --))
)
(GRAFMCMETHYL
         (1 C X (2 3 4) 1 0)
         (2 -- X (1 --) 1)
         (4 -- X (1 5 --))
         (5 -- X (4 6 --))
         (6 C X (5 7 --))
         (7 -- X (6 8 --))
         (8 -- X (7 --))
         (3 -- X (1 9 --))
         (9 -- X (3 10 --))
         (10 C X (9) 0 3)
)
(GRAFDBLMCALPHA
         (1 C X (2 3 4) 1)
         (2 -- X (1 --) 1)
         (4 -- X (1 5 --))
         (5 -- X (4 6 --))
         (6 C X (5 --))
         (3 -- X (1 9 --))
         (9 C X (3 7 10) 0 0)
         (7 C X (9 8 --))
         (8 C X (7 --))
         (10 C X (9 --))
)
(GRAFETH1
         (1 C X (2 --))
         (2 (O N S) X (1 3 --))
         (3 C X (2 4 --))
         (4 C X (3 --))
)
(GRAFCH3ALPHA
         (1 C X (2) 0 3)
         (2 (N O) X (1 --))
)
(GRAFHAL1*3
         (2 (FL CL BR I) X (1) 0 0)
         (1 C X (2 3 --))
         (3 C X (1 --))
)
(GRAFHAL1*4
         (2 (FL CL BR I) X (1) 0 0)
         (1 C X (2 4 --))
         (4 C X (1 3 --))
         (3 C X (4 --))
)
```

Table 6-1 (*Continued*)

```
(GRAFH2O1*2
    (2 O X (1) 0 1)
    (1 C X (2 3 --))
    (3 C X (1 --))
)
(GRAFH2O1*3
    (2 O X (1) 0 1)
    (1 C X (2 4 --))
    (4 C X (1 3 --))
    (3 C X (4 --))
)
(GRAFH2O1*4
    (2 O X (1) 0 1)
    (1 C X (2 4 --))
    (4 C X (1 5 --))
    (5 C X (3 4 --))
    (3 C X (5 --))
)
(GRAFH2S1*3
    (2 S X (1) 0 1)
    (1 C X (2 3 --))
    (3 C X (1 --))
)
(GRAFH2S1*4
    (2 S X (1) 0 1)
    (1 C X (2 4 --))
    (4 C X (1 3 --))
    (3 C X (4 --))
)
(GRAFNH31*3
    (2 N X (1) 0 2)
    (1 C X (2 3 --))
    (3 C X (1 --))
)
(GRAFNH31*4
    (2 N X (1) 0 2)
    (1 C X (2 4 --))
    (4 C X (1 3 --))
    (3 C X (4 --))
)
(GRAFCH41*3
    (2 C X (1) 0 3)
    (1 C X (2 3 --))
    (3 C X (1 --))
)
(GRAFCH41*4
    (2 C X (1) 0 3)
    (1 C X (2 4 --))
    (4 C X (1 3 --))
    (3 C X (4 --))
)
(GRAFMC5
    (1 C X (2 4 --) 1)
    (2 (O N C) X (1 --) 1)
    (4 (C N O) X (1 5 --))
    (5)(C N O) X (5 6 --))
    (6 C X (5 7 --))
    (7 C X (6 8 --))
    (8 C X (7 --))
)
(GRAFCO
    (1 C X (2 3) 2 0)
    (2 O (1) 1 0)
    (3 C X (1 --))
)
```

COELIM

1. Break the bond between nodes 2 and 3.
2. Return a list containing the ion containing node 3.

COMPL

Place the complement of the previous ion directly into the mass spectrum. Return no ions.

DELETEPARENT

Reduce the intensity of the parent ion to 0. Return no ions.

ELIM

1. Remove a hydrogen from node 3.
2. Break the bond between nodes 1 and 2.
3. Return a list containing the ion containing node 1.

GAMMACLEAVAGES

1. Find all bonds gamma to node 2.
2. Break each bond in turn.
3. For each bond broken, create an ion containing node 2.
4. Return the list of ions created.

LASTION

Return a list containing the previous ion.

LOSE1*2

1. Break the bond between nodes 1 and 3.
2. Return a list containing the ion containing node 3.

MAJORALPHACLEAVAGE

1. Find all major bonds alpha to node 2.
2. Break each bond in turn; and for each bond broken, create an ion containing node 2.
3. Return the list of ions created.

MINORALPHACLEAVAGES

1. Find all bonds alpha to node 2 that are not major bonds.
2. Break each bond in turn; and for each bond broken, create an ion containing node 2.
3. Return the list of ions created.

MRRFGT

1. Migrate a hydrogen atom from node 6 to node 2.
2. Break the bond between nodes 4 and 5.
3. Return a list containing the ion containing node 4.

MRRFGT2

1. Migrate a hydrogen atom from node 8 to node 2.
2. Break the bond between nodes 3 and 9.
3. Return a list containing the ion containing node 3.

MRRFGT3

1. Migrate a hydrogen atom from node 10 to node 2.
2. Break the bond between nodes 3 and 9.
3. Return a list containing the ion containing node 3.

MRRX (AT1, AT2, AT3)

1. Migrate a hydrogen atom from node AT1 to node 2.
2. Break the bond between nodes AT2 and AT3.
3. Return a list containing the ion containing node AT2.

PERFORMBREAK (BND); a function

args: BND; a bond (dotted pair of active-node descriptors).
Value: the ion created by breaking the bond and keeping the fragment attached to the first node.
Comment: this function is not intended for use as a theory-statement primitive.

REMOVEH1

1. Eliminate a hydrogen atom from node 1.
2. Return a list containing the ion containing node 1.

INTENSITY FUNCTIONS

LASTINT (N)

((intensity of previous ion) × N)/100

MAJORALPHAINT

(intensity of the parent ion) × KCMAJ

MINORALPHAINT

M1 ← (carbon count of the ion being computed)
M2 ← (carbon count of the parent ion) − M1 − 1
T1 ← (intensity of the previous ion) × M1
if M2 = 0 then T1/KCMAJ
           else T1/(KCMAJ × M2)

6.3.3 An Example

To illustrate the output of PREDICTOR we take the now-familiar example of an estrogen, estradiol, and have the program predict its low-resolution spectrum and

HEURISTIC DENDRAL TESTING

metastable peaks. The production for estrogens, given above, was used in this run. Recall that the intensity computations used simply copied the current intensity to each ion generated. Thus the spectrum produced has peaks of equal height. This result is not inherent in the program, which can use arbitrarily complex computations, but it simplifies this example.

The input to the program is the structure defining estradiol (the chemical graph is as depicted in Figure 5-4, with OH groups added to nodes 3 and 17):

```
ESTRADIOL
((ESTRADIOL) (CHEMICAL
 (C1  C  1  (C10 C2)     1 1)
 (C2  C  2  (C3 C1)      1 1)
 (C3  C  3  (O19 C4 C2)  1 0)
 (C4  C  4  (C5 C3)      1 1)
 (C5  C  5  (C10 C6 C4)  1 0)
 (C6  C  6  (C7 C5)      0 2)
 (C7  C  7  (C8 C6)      0 2)
 (C8  C  8  (C14 C9 C7)  0 1)
 (C9  C  9  (C11 C10 C8) 0 1)
 (C10 C 10  (C5 C1 C9)   1 0)
 (C11 C 11  (C12 C9)     0 2)
 (C12 C 12  (C13 C11)    0 2)
 (C13 C 13  (C17 C18 C14 C12) 0 0)
 (C14 C 14  (C8 C15 C13) 0 1)
 (C15 C 15  (C16 C14)    0 2)
 (C16 C 16  (C17 C15)    0 2)
 (C17 C 17  (O20 C13 C16) 1 0)
 (C18 C 18  (C13)        0 3)
 (O19 O 19  (C3)         0 1)
 (O20 O 20  (C17)        0 1)))
STOP
```

The output of the program is as follows, annotated where appropriate.

PREDICTOR CONSTRAINTS for ESTRADIOL

(PARAMETER NAME)	VALUE
(HIPEAKNORMFLAG)	T [*Note:* T(rue) means the highest peak will be assigned intensity 100 and the others will be normalized to this level; false means the intensities will be made to sum to 100.]

show only masses in the spectrum? Y/N.
The usual response is: N

(ONLYMASSES)	NIL [*Note:* only masses means no intensities.]

maximum number of steps in a fragmentation process.

The usual response is: 2
   (MAXIONLEVEL)   2
spectrum resolution? HIGH/LOW.
   (RESOLUTION)   LOW

END.

REVIEW OF MASS SPECTRUM PREDICTION for ESTRADIOL

	BREAK	CHARGE	COMPOSITION	MASS	METASTABLE PEAK
ION# 0	M+	10	C18H23O2	271	
ION# 1	B	10	C15H18O	214	M*PEAK 169
ION# 1	B	10	C15H17O	213	M*PEAK 167
ION# 2	D	10	C13H14O	186	M*PEAK 128
ION# 2	D	10	C13H13O	185	M*PEAK 126
ION# 3	C	10	C12H13O	173	M*PEAK 110
ION# 3	C	10	C12H12O	172	M*PEAK 109
ION# 4	E	10	C11H12O	160	M*PEAK 94
ION# 4	E	10	C11H11O	159	M*PEAK 93
ION# 5	F	10	C10H10O	146	M*PEAK 79
ION# 5	F	10	C10H9O	145	M*PEAK 78

END.

6.4 RANKING THE CANDIDATE EXPLANATIONS

Different models, or "theories," of mass spectrometry can be used to predict the fragmentations of a molecular skeleton. The type of fragmentation theory to be used depends largely on the context of the structure determination problem. When one initially studies a new class of compounds, or when one attempts to discriminate among different candidate structures obtained from some unusual CONGEN problem, it is usually appropriate to use some universal form of fragmentation theory that expresses very general chemical principles. Although a general theory will be applicable and will not be biased in its predictions, it may well prove to have poor discriminatory power. Fine discrimination between related structures generally requires more refined fragmentation theories wherein one assigns different plausibilities to alternative fragmentation processes. When processing isomers from some well-characterized class, the appropriate fragmentation theory may well involve the detailed specification of substructures, their bond cleavage processes, and the accompanying specific transfers of hydrogen atoms or other molecular fragments.

In the following subsections we show how even the most general "half-order" fragmentation theory can serve to discriminate among isomers of moderately complex structures such as monoketoandrostanes. (The skeleton of this class is illustrated in Figure 7-2.) Refinements of the simplest half-order theory involve first the use of estimates of relative plausibilities of fragmentation processes of differing degrees of complexity and, subsequently, the association of relative plausibility values with some classes of bond cleavage. The use of detailed class-specific fragmentation processes is considered in relation to the processing of the spectra of macrolide antibiotics.

6.4.1 The Half-Order Theory of Molecular Fragmentation

The simplest model of molecular fragmentation is the ALLBREAKS, or zero-order, "theory" that predicts ions arising from all bond cleavages and combinations of cleavages and transfers of atoms between fragments. Such a model is too general for almost every problem in computer analysis of mass spectra. (However, the zero-order theory was the method of spectrum prediction in the first application of Heuristic DENDRAL to amino acids.) DENDRAL's "half-order theory" of mass spectrometry is a constrained version of the ALLBREAKS model of molecular fragmentation: of all the possible fragmentations, some are not tried because they are implausible, if not patently absurd. The half-order theory is a very loose model; it does not describe the detailed aspects of fragmentations, such as specific relationships between atom transfers and cleavage of certain groups of bonds, and it makes no attempt to express anything about the mechanisms by which fragmentations actually take place.

The constraints that can be expressed in the half-order theory of molecular fragmentation include limitations on the number of bonds that may be broken and the number of allowed hydrogen transfers into or out of the charged fragment. Each predicted ion is formed by a "process" involving (1) one or more cleavage "steps," (2) possible H-transfers, and (3) possible neutral losses. Each "step" cleaves a molecule and may be a break of one acyclic bond, two bonds within a ring, or a group of three bonds in an edge-fused-ring system. A complete process could, for example, involve fused-ring cleavage, simple ring cleavage, and acyclic bond cleavage steps; such a process would involve a total of six bond breaks. Typical constraints used with the half-order theory would be:

1. Prohibit cleavage of aromatic or isolated double or triple bonds.
2. Allow one or two step processes.
3. Allow at most two bonds to be cleaved in a given step.
4. Permit a maximum of three bonds to be cleaved in a process.
5. Prohibit the cleavage of two (nonhydrogen) bonds from the same carbon atom (for this cleavage would formally leave a fragment that is normally energetically unfavorable).
6. Restrict transfers between fragments to at most two hydrogen atoms.

A simple use of the half-order theory for testing is implemented in the MSPRUNE function. MSPRUNE helps a chemist reject CONGEN structures by determining the difficulty of rationalizing any specific ion in the spectrum from each of the possible structures. Even in this very limited form, the half-order theory can be of value in helping to identify candidate structures compatible with spectral data. For example, a simple ring cleavage and hydrogen transfer are a simpler explanation than cleavage of a fused-ring system. MSPRUNE uses such differences to eliminate candidate structures [Smith and Carhart (1978)].

Table 6-2 Ranking of monoketoandrostanes based on the half-order theory

Structure (keto position)	Ranking	Structures with equal score	Better-ranked structures
1	1	2, 3, 4	
3	1	1, 2, 4	
4	1	1, 2, 3	
6	1		
7	2		6
11	2		12
12	1		
15	1	17, 16	
16	1	17, 15	
17	2	15, 16	1, 2, 3, 4

Note: Fragmentation constraints: one-step processes, a maximum of two bonds cleaved, transfer of at most two hydrogens into or out of the charged fragment.

Generally, we have found it to be more effective to employ the data in the entire observed mass spectrum and rank candidate structures according to how well they serve to explain the spectral data. This ranking is accomplished through the MSRANK function, which allows the user to define the constraints of the half-order theory and to specify the form of the scoring function. The score assigned to a candidate structure is determined from the importance accorded to those of the observed ions that can be generated by the allowed fragmentations of that candidate. As ions at higher mass and intensity values are generally of greater structural significance, the importance accorded to each observed ion in the spectrum is determined by some function of its m/e and its relative intensity (in most cases, the product of m/e and intensity has been used).

The results shown in Table 6-2 typify the performance of this simple approach to discriminating between structures. The structures analyzed were monoketoandrostanes, all with the same steroid skeleton but varying in the position of the keto substituent. The simple half-order-theory approach was used, and the 11 possible isomers were ranked against each of the 10 available high-resolution mass spectra. The half-order theory generally separates structures into two groups, those that match the spectra about equally well and those that can definitely be eliminated. With these structures, the correct candidate was generally ranked first after its predicted and recorded spectra had been compared, but it was not possible to discriminate among isomers with the keto group on nodes 1 to 4 or among those with the keto group on nodes 15 to 17.

6.4.2 Half-Order Theory with Process and Bond-Break Plausibilities

Differences among candidate structures are not always simply explained by the major fragmentation processes. In such cases, we can use the MSRANK program with a more refined version of the half-order theory in which relative plausibility values, in the range 0 to 1, are associated with processes according to differing numbers of steps, differing numbers of bond cleavages, and different types of neutral transfers between fragments. The principle behind this reasoning is that if two structures both provide explanations for an observed ion, then the structure with the simpler explanation is more likely. The plausibility of a predicted ion is given as the product of the break plausibilities of the bonds and H-transfers or neutral losses involved, modified by any additional factors such as the reduced plausibility of a process requiring adjacent breaks or multiple steps. If an observed ion can be rationalized in terms of two different fragmentations of the same structures, then the process with the higher plausibility is used. The score assigned to a structure is the sum, over all observed ions, of the importance of the ion (m/e times intensity) multiplied by the plausibility of the process producing it. With these plausibility weighting factors, the half-order theory can discriminate quite well between related structures. Typically, around half or even two-thirds of a set of candidate isomers can be rejected on the results of mass spectral ranking.

6.4.3 The Use of Class-Specific Fragmentation Rules in MSRANK

The half-order theory typically fails to discriminate within a large group of equally well-ranked structures even when relative plausibilities of different fragmentations are defined. This failure is partly due to the fact that the candidates will have very similar structures. However, the overgenerality of the theory also contributes to this degeneracy of scores. The half-order theory does not allow specific hydrogen transfers or neutral losses to be associated with specific break processes.

In structures with several charge-localization/fragmentation-directing substituents, interactions between competing fragmentation processes must be expected. It is hard to predict a priori the result of such interactions on the appearance of the mass spectrum, and general half-order theories can be of limited value. However, if the candidate structures are from a previously studied class, then rules defining their fragmentation behavior can supplement, or replace, the half-order theory to explain observed ions and rank structures. Generally, class-specific fragmentation rules produce a finer discrimination among candidates.

The rules given to the program must specify substructures, bond breaks, and specific transfers. Such rules have been derived manually, by chemists working with DENDRAL, for a number of compound classes including the macrolide antibiotics. Some of the standard macrolides are shown in Figure 6-1. Conventional analysis (and results from the INTSUM program, described later in Chapter 7) showed that the major fragmentations of the macrolide skeleton could be described in terms of certain McLafferty rearrangements, cleavages alpha to carbonyl groups, and other processes illustrated in Figure 6-2. To test the discriminatory power of the fragmentation rules that had been derived, isomers of the standard macrolactones were generated using CONGEN. In each case the standard macrolactone skeleton was retained, the isomers varying only in the position of hydroxy, keto, and alkyl substituents and in the position of the double bond in the macrolactone ring. The generated isomers were ranked using the experimental mass spectrum of the standard compound. If just the half-order theory, without any plausibility weightings was used, the spectrum-ranking func-

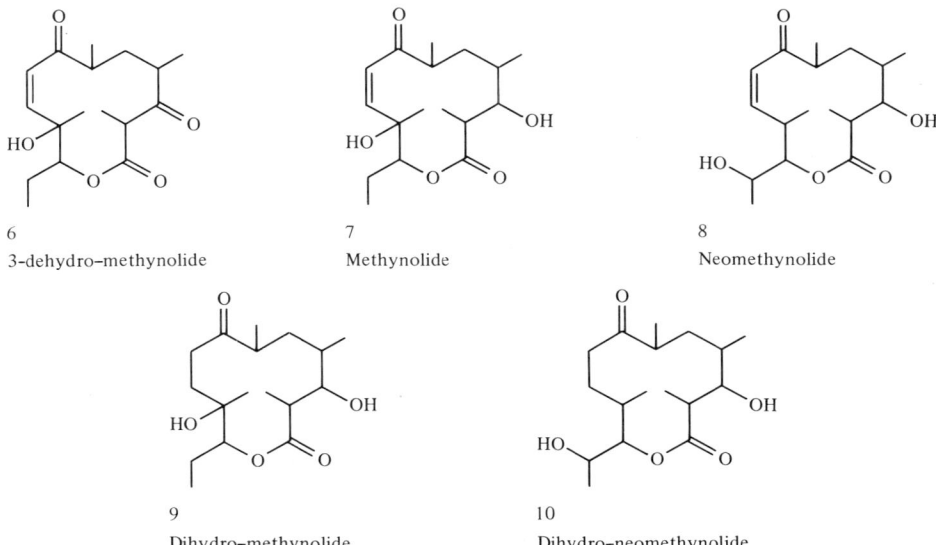

Figure 6-1 Some standard macrolides. (*Source:* Biochemical Applications of Mass Spectrometry (supplement), *edited by G. R. Waller. Copyright 1979, John Wiley & Sons, Inc. Reproduced by permission. After Gray et al. [1979].*)

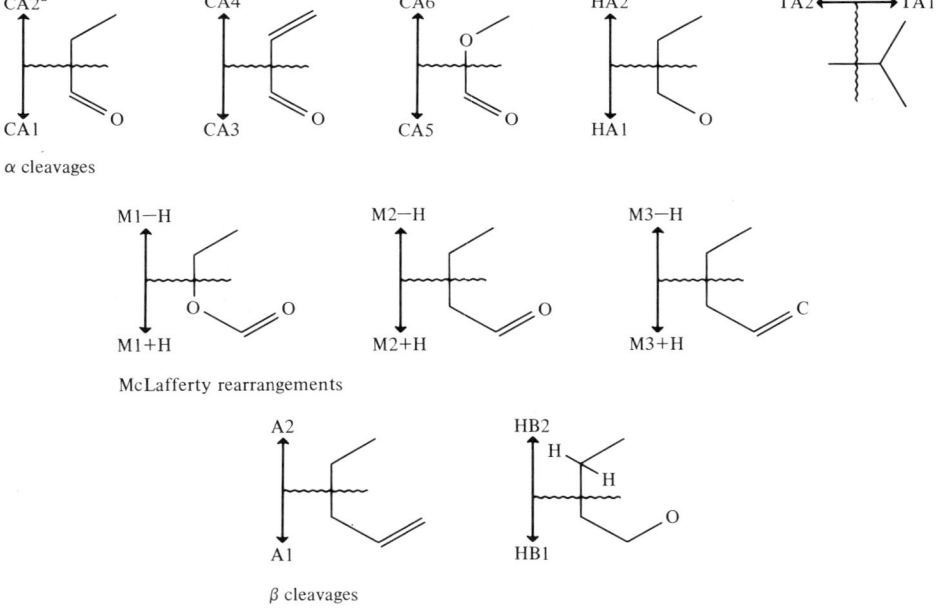

Figure 6-2 Rules used in predicting the mass spectra for macrolide structures. (*Source:* Biochemical Applications of Mass Spectrometry (supplement), *edited by G. R. Waller. Copyright 1979, John Wiley & Sons, Inc. Reproduced by permission. After Gray et al. [1979].*)

104

Table 6-3 Comparison of performance of MS-ranking functions when using half-order theory and class-specific rules

Structure	Total # of isomers	Number of isomers with equally good explanations of recorded spectrum	
		Half-order theory	Class-specific rules
6	60	5	2
7	105	28	1
8	105	4	2
9	105	27	5
10	105	10	5

Note: The structures are the macrolide antibiotics shown in Fig. 6-1. Sets of CONGEN-generated isomers were ranked using the recorded high-resolution mass spectrum of the standard structures.

tions could always eliminate at least 75 percent of the isomers. However, the correct structure was never uniquely identified.

The rules that had been derived for these skeletons involved all pairwise combinations of the single-break processes shown in Figure 6-2. Specific hydrogen transfers were associated with each rule. As shown in Table 6-3, use of the rules in the spectrum-generating process resulted in greatly reduced ambiguity. With rules, methynolide could apparently be identified with certainty, and, typically fewer than 1 isomer in 20 was found to be compatible with the spectral data.

6.5 SUMMARY OF HEURISTIC DENDRAL

The Heuristic DENDRAL system consists of planning, generating, and testing phases. Each may be used alone, or the phases may be used together. The programs themselves embody theories of chemical structure and mass spectrometry. Particular hypotheses for particular classes of compounds are formulated by the user and saved. The knowledge thus formulated gives the program its power.

1. Planning
    a. Planning may use special heuristic rules relating sets of mass spectrum peaks with subgraphs of specific compound classes, as was done in early versions for aliphatic compounds.
    b. A Planning Rule Generator may be used to systematically consider all possible fragmentation processes of specified type. It writes the planning rules relating peaks to subgraphs.
    c. The DENDRAL PLANNER may be used to produce automatically constraints from user-supplied definitions of break patterns for a known class of compounds. It subsumes the Planning Rule Generator.

    d. MOLION may be used alone or as part of PLANNER to infer the molecular ion from a mass spectrum.
    e. Manually inferred constraints may be supplied to CONGEN.
2. Generating
CONGEN is a generator of a complete and nonredundant list of isomers for a given empirical formula, including structures containing rings. It subsumes the original acyclic generator for aliphatic compounds. Its current power comes from user-supplied constraints, although constraints produced by PLANNER may also aid in narrowing the output of the generating phase.
3. Testing
    a. The PREDICTOR is a production-system-oriented program that applies fragmentation rules to a given structure to produce a mass spectrum.
    b. The predicted spectrum may then be compared to data to determine the plausibility of each candidate structure as a source of the data.

CHAPTER
SEVEN

META-DENDRAL

Meta-DENDRAL is a separate program that discovers new rules of mass spectrometry and ^{13}C NMR spectrometry for use by the planning and testing programs of Heuristic DENDRAL. It is more of an induction program than Heuristic DENDRAL in the traditional sense of proposing general rules that explain a number of observed instances. Nevertheless, the plan-generate-test paradigm is used by Meta-DENDRAL as well. The key idea is to search a space of possible rules using both the empirical data and a strong model of the domain to guide the search.

7.1 INTRODUCTION

We have seen that Heuristic DENDRAL employs user-supplied knowledge to guide its search through a space of possible connectivity isomers to produce a list of isomers that are likely explanations of the given mass spectrum (or other data). The compilation and translation of the heuristic knowledge into the codification appropriate to Heuristic DENDRAL requires anywhere from a few weeks to a few months of the time of a chemist who is familiar not only with the subject class of compounds and its behavior in the mass spectrometer, but with DENDRAL as well.

One significant way in which the system can be generalized is to automate this *knowledge acquisition* phase. It may be hoped that such a generalization will do more than facilitate the application of DENDRAL to new compounds. A careful study of the processes of induction may lead to new insights about the theory of mass spectrometry, i.e., how molecular structure determines the behavior of molecules under electron bombardment. It may also lead to a deeper understanding of the means by which scientists acquire and use specialized knowledge. With these objectives in mind, a substantial effort has been devoted to the development of a knowledge acquisition system called Meta-DENDRAL.

It is useful to distinguish several approaches that might be taken toward the realization of a system for automatic knowledge acquisition. Five schemes are outlined in Figure 7-1.

In the first, which we have called custom crafting, the route from knowledge source to program is traversed in two stages. The information source is interpreted by a scientist, and a programmer, studying the scientist's report of his activity, codifies the information (the scientist's knowledge) in a manner suitable for machine simulation. This method is the one used in the construction of early versions of Heuristic DENDRAL [see in particular Buchanan, Sutherland, and Feigenbaum (1970)]. As Heuristic DENDRAL evolved further, knowledge acquisition was effected by the second scheme in which a single interlocutor (e.g., Raymond Carhart) was knowledgeable in both chemistry and the Heuristic DENDRAL system. Much of AI research has used this approach, usually because AI researchers were programmers and tended to choose a subject matter in which they were expert also. For example, chess, natural language, logic, and mathematics have been subjects of much investigation largely for this reason. A more sophisticated, or at least more automatic, approach is what we have called the dialogue approach. This approach requires a program that is sufficiently competent to assume some of the work done by the "programmer" in the previous cases, thereby enabling it to work with scientists who do not have intimate familiarity with the program, or with programming. Here we envision a program that listens to scientists' descriptions of their methods and asks them pertinent questions. The TEIRESIAS program constructed by Davis (1976) is an example of this type of program.

In the last two approaches diagrammed, the human scientist and/or programmer have been eliminated, and the program is able to examine the information source directly. We have distinguished two major classes of information sources, text and data.[1] An automatic text analyzer would, ideally, be able to "pick up" a textbook, read and interpret it, and augment its own knowledge base. To our knowledge no existing system even approaches such an ability, though the concept has often been considered. The final approach, in which the program directly examines data and induces generalizations about regularities, is the approach taken by the Meta-DENDRAL research. Note that "data" does not refer to raw observations of the world via human-

---

[1] A similar distinction can be made in the different sources used by scientists in the custom-crafting model, which requires different kinds of reasoning by the scientist.

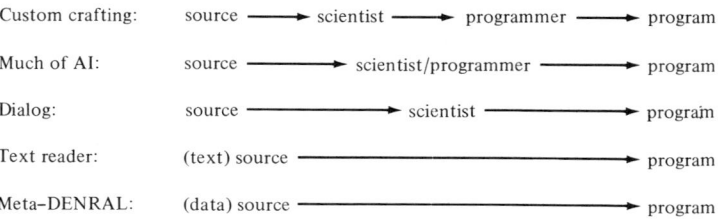

Figure 7-1 Approaches to automatic knowledge acquisition.

like sense organs, but rather to already structured (and quantified) measurements, in particular, mass spectra.

What Meta-DENDRAL induces is a set of productions (situation-action rules) that constitute a partial theory of fragmentation for a class of compounds. These productions are the same sorts of rules as those used by Heuristic DENDRAL in planning and testing. The productions are induced in a three-stage procedure that is a species of the plan-generate-test paradigm.

The stages of Meta-DENDRAL's knowledge acquisition are:

1. The INTSUM (INTerpretation and SUMmary) program examines a collection of spectra from several members of a class of compounds and seeks to account for the important peaks in terms of *MS processes*. A process specifies (a) bond cleavages, (b) the charged fragment, (c) a resultant composition of the charged fragment, and (d) a set of associated migrations (transfers) of hydrogen atoms and electrically neutral fragments.
2. RULEGENerator represents these processes in terms of productions describing the general environments in which the processes apply.
3. RULEMODifier refines this set of productions by generalization, selection, merging, and specialization.

7.2 INTSUM—DATA INTERPRETATION AND SUMMARY

This program requires that the user supply a skeleton structure common to the members of the class. The skeleton may be a single atom, e.g., the nitrogen atom common to the aliphatic amines, or a large ring structure, such as the steroid nucleus that has been used as an example in previous chapters. The skeleton is defined and coded in the now-familiar manner.

The user may next define part of the semantic model of mass spectrometry that guides the whole system, by specifying kinds of fragmentation processes to be considered. This definition is done by specifying constraints on the types of processes in a so-called half-order theory of mass spectrometry. If no constraints are specified, i.e., the zero-order theory of mass spectrometry is used, all possible MS processes (combinations of bond fragmentations and neutral transfers) are considered. Usually, however, the user will wish to shorten the search by limiting the process types in a number of ways. The types of constraints are the following options:

1. Cleave (or not) more than one bond to a single atom prohibited.
2. Cleave (or not) aromatic ring bonds.
3. Cleave (or not) double and triple bonds.
4. Retain a minimum number of carbon atoms in a fragment.
5. Accompany cleavage with a range of allowed hydrogen transfers (numbers and directions).
6. Cleave a maximum number of bonds in a single process.

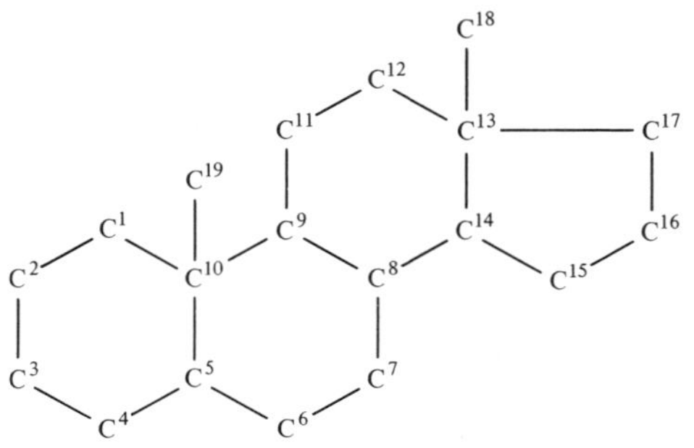

Figure 7-2 Androstane skeleton.

7. Allow a maximum number of steps in a process. (Two steps would allow fragmentation of fragments, etc.)
8. If multiple step processes have been allowed, cleave a maximum total number of bonds.
9. If multiple rings have been allowed to break, cleave a maximum number of rings.
10. Permit neutral fragments other than hydrogen to transfer.

This model defines, implicitly, a set of permitted MS processes for the given skeleton. A report summarizing this basic information is printed. Each process is given a name, called a process label, and is identified by a list of bonds (that cleave), an indicator of which fragment is charged, and a fragment composition (giving the composition of the charged fragment). At this point only the skeleton has been considered, and the neutral transfer information has not yet been used.

One class of compounds to which Meta-DENDRAL has been applied is the monoketoandrostanes, another subclass of steroids. The skeleton of this class is depicted in Figure 7-2. There are 11 possible monoketoandrostanes corresponding to the 11 possible carbon atoms in the rings to which an oxygen atom can be double bonded (recall that a keto group is C=O): nodes 1, 2, 3, 4, 6, 7, 11, 12, 15, 16, and 17. The analysis here described was based on spectra from 10 of these, plus spectra from two stereoisomers and from androstane (whose structure is just the skeleton), making 13 molecules and spectra in all. A portion of the output file produced by INTSUM as the first step of its analysis follows. The parameter values listed define the model that the user supplied for this run.

FRAGMENTATION PROCESS CONSTRAINTS for ANDROSTANE

(PARAMETER NAME)   VALUE
----
1.   forbid cleavage of more than one bond to the same atom?
        the usual response is: Y
              (ADJBONDFILTERFLAG)   T

2.   forbid cleavage of aromatic ring bonds? Y/N.
        the usual response is: Y
              (AROMATICFILTERFLAG)   T

#	(PARAMETER NAME)	VALUE
3.	forbid cleavage of double and triple bonds? Y/N.	
	the usual response is: Y	
	(BONDFILTERLIST)	(DOUBLE TRIPLE)
4.	minimum number of carbons in a fragment?	
	the usual response is: 0	
	(MINCARBONFILTER)	2
5.	allowed hydrogen transfers?	
	the usual response is: 0	
	(HTRANSFERS)	(-2 -1 0 1 2)
6.	maximum number of bonds allowed to cleave in a single step process?	
	the usual response is: 4	
	(MAXBREAKORDER)	2
7.	maximum number of steps in a fragmentation process?	
	the usual response is: 2	
	(MAXPROCESSLEVEL)	1
8.	maximum number of bonds allowed to cleave in a multiple step process?	
	the usual response is: 6	
	(MAXBONDBREAKS)	6
9.	maximum number of rings allowed to fragment in a multiple step process?	
	the usual response is: 3	
	(MAXRINGBREAKS)	3
10.	allowed neutral transfers (other than H)?	
	(TRANSFERS)	NIL

END.

SKELETON DRAWING for ANDROSTANE

```
    2- 1     11-12
   /   \    /    \1
  3   19-1-9   18-3-17
   \   /0  \    /    |
    4-5     8-14     |
     \   /    \      |
      6-7     15-16
```

ALL ATOMS ARE CARBON.

LIST OF PROCESSES DEFINED for ANDROSTANE

37 PROCESS LABELS DEFINED
(THE DEFAULT SET OF H-TRANSFER VALUES IS) (-2 -1 0 1 2)

PROCESS LABEL	FRAGMENT SELECTOR	FRAGMENT COMPOSITION	LIST OF BONDS
BRK0	19	C19H32	
BRK1L	10	C18H29	(10 19)
BRK2L	13	C18H29	(13 18)
BRK3L	1	C17H28	(1 2) (3 4)
BRK3H	2	C2H4	(1 2) (3 4)
BRK4L	1	C16H26	(1 2) (4 5)
BRK4H	2	C3H6	(1 2) (4 5)
BRK5L	1	C2H4	(1 10) (2 3)
BRK5H	10	C17H28	(1 10) (2 3)
BRK6L	1	C3H6	(1 10) (3 4)
BRK6H	10	C16H26	(1 10) (3 4)
BRK7L	1	C4H8	(1 10) (4 5)
BRK7H	10	C15H24	(1 10) (4 5)
BRK8L	2	C17H28	(2 3) (4 5)
BRK8H	3	C2H4	(2 3) (4 5)
BRK9L	5	C17H28	(5 6) (7 8)
BRK9H	6	C2H4	(5 6) (7 8)
BRK10L	5	C7H12	(5 6) (9 10)
BRK10H	6	C12H20	(5 6) (9 10)
BRK11L	6	C8H14	(6 7) (9 10)
BRK11H	7	C11H18	(6 7) (9 10)
BRK12L	7	C9H16	(7 8) (9 10)
BRK12H	8	C10H16	(7 8) (9 10)
BRK13L	8	C11H18	(8 14) (9 11)

PROCESS LABEL	FRAGMENT SELECTOR	FRAGMENT COMPOSITION	LIST OF BONDS
BRK13H	14	C8H14	(8 14) (9 11)
BRK14L	8	C12H20	(8 14) (11 12)
BRK14H	14	C7H12	(8 14) (11 12)
BRK15L	8	C13H22	(8 14) (12 13)
BRK15H	14	C6H10	(8 14) (12 13)
BRK16L	9	C17H28	(9 11) (12 13)
BRK16H	11	C2H4	(9 11) (12 13)
BRK17L	13	C16H26	(13 17) (14 15)
BRK17H	17	C3H6	(13 17) (14 15)
BRK18L	13	C17H28	(13 17) (15 16)
BRK18H	17	C2H4	(13 17) (15 16)
BRK19L	14	C17H28	(14 15) (16 17)
BRK19H	15	C2H4	(14 15) (16 17)

END.

Following the construction of the list of basic skeletal MS processes, INTSUM is ready to examine spectra. Each spectrum is examined in turn, and significant peak groups are determined by the same local weighting scheme used by MOLION (Section 5.3) in order that important high mass peaks will not be ignored because of their low abundances. At this point, fragmentations in the substituents of each molecule are also defined and added to the list of basic processes to be considered for that molecule. For each of the significant peak groups, an exhaustive search of the list of processes is made to locate all processes that yield a charged fragment of the appropriate composition.

A summary of findings is printed for each spectrum. Each peak for which an explanation was found has an abundance that may be expressed as a percentage of the total ion current. The sum of these intensity percentages is an index of how much of the important data has been accounted for by the listed processes. The following is an example of the results for a spectrum for just one monoketoandrostane. Process names refer to labels listed in the previous table. Transfers when proposed are indicated by concatenating a colon to the process label, followed by the direction and number of hydrogens that migrate. (Minus indicates loss, or transfer out of the charged fragment; plus indicates gain, or transfer into the charged fragment.) Other atoms or neutral fragments such as H_2O may also be specified.

MASS SPECTRUM for S:ANDROSTAN-3-ONE-MAPA1204

(SKELETON IS)   ANDROSTANE
(SUBSTITUENTS ARE)   ((SUBSTO= 3))
(MOLECULE NUMBER)   4
(TOTAL ION CURRENT IS)   1162
3 (PEAKS BELOW MINIMUM MASS OR INTENSITY)
85 (PEAKS REMOVED BY CLUSTERING)   [See Section 5.3.1.]
(HAVING A COMBINED INTENSITY OF)   21.1

END.

[Note that the keto group is at node 3, i.e., the oxygen atom (node 20) is doubly bonded to the carbon atom at node 3, as represented by the dots at 3 and 20.]

```
       12-11     1-2
        1/  \   /  \ 2
   17-3-18  9-1-19 3.-0.
    |   \   / 0\  /
    |   14-8    5-4
    | /    \  /
   16-15    7-6
```

NON-CARBON ATOMS ARE: (O 20)

PEAK EXPLANATIONS for S:ANDROSTAN-3-ONE-MAPA1204

MOLECULE S:ANDROSTAN-3-ONE-MAPA1204

	(DATA POINTS)		(PROPOSED EXPLANATION)	
	FRAGMENT	INTENSITY		
MASS	COMPOSITION	(% TOT ION)	PROCESS NAMES	
274	C19H30O	6.05	BRK0	
259	C18H27O	1.58	BRK1L	BRK2L

[$(M^+ - 15)$ can be a loss of either methyl group. Thus there are two possible explanations.]

232	C16H24O	.55	BRK17L	
231	C16H23O	1.49	BRK17L:-H	
219	C16H27	.46	BRK4L:+H	BRK6H:+H
217	C16H25	.59	BRK4L:-H	BRK6H:-H
216	C16H24	.25	BRK4L:-2H	BRK6H:-2H
203	C15H23	5.14	BRK7H:-H	
202	C15H22	8.51	BRK7H:-2H	

177	C12H17O	.20	BRK14L:-H
163	C12H19	.46	BRK10H:-H
163	C11H15O	.53	BRK13L:-H
162	C12H18	.34	BRK10H:-2H
149	C11H17	1.38	BRK11H:-H
148	C11H16	.64	BRK11H:-2H
135	C10H15	1.49	BRK12H:-H
124	C8H12O	.84	BRK11L
123	C8H11O	.56	BRK11L:-H
109	C8H13	1.78	BRK13H:-H
108	C8H12	1.28	BRK13H:-2H
95	C7H11	5.06	BRK14H:-H
81	C6H9	3.64	BRK15H:-H
41	C3H5	3.54	BRK17H:-H

(TOTAL INTENSITY OF ALL DATA POINTS SHOWN IS) 46.3

(THOSE DATA POINTS NOT HAVING ANY EXPLANATION)
-----
25 (DATA POINTS WITH INTENSITY ABOVE) 0
(HAVING A COMBINED INTENSITY OF) 32.5

END.
(RESULTS SAVED FOR) ANDROSTANE

Note that roughly one-third of the total ion current remains unexplained under this model, which was specified by the chemist. At this point the chemist may wish to change the model and run INTSUM again if there are significant peaks in the spectrum that must be explained. The half-order theory may also be augmented with specific processes (however complex) that the chemist chooses to include in the model.

The final output from INTSUM is an INTSUM *table of information* about each break giving break name, the names of the molecules for which there is *positive evidence* (predicted peaks that are found in spectra) for this break, and the average intensity of the ion current of positive evidence for each of the molecules. The INTSUM table for androstanes is shown in Table 7-1.

The INTSUM table of results is itself an informative and useful codification of a set of mass spectra. Because the model of mass spectrometry can be defined to capture the chemist's expectations, INTSUM can quickly point out the unexpected peaks. In addition, plausible alternative explanations are often surprising in themselves, since chemists do not usually exhaust the combinations of allowable fragmentations and transfers.

7.3 RULEGENeration

As defined by INTSUM, each process is specific to particular bonds, the break loci. For example, BRK3L cleaves the carbon-carbon bonds between nodes (1 2) and (3 4)

Table 7-1 Part of the INTSUM table for monoketoandrostanes

REPORT OF ALL PROCESSES HAVING EVIDENCE for ANDROSTANE				
31 SEPARATE PROCESS LABELS				
FULL PROCESS LABEL	OCCURRENCE RATIO (MOLS SHOWING THIS PROCESS)	MOL ID	INTEN SITY	PARTIAL REDUNDANCIES
BRK0	13/13	1	11.5	
		2	11.3	
		5	9.94	
		9	9.17	
		6	8.75	
		13	8.19	
		3	7.82	
		10	7.69	
		8	7.05	
		12	6.08	
		4	6.05	
		7	5.78	
		11	5.27	
			8.05	(AVERAGE INTENSITY)
BRK1L ((10 19))	13/13	7	9.03	BRK2L
		6	7.87	BRK2L
		9	5.14	BRK2L
		10	4.23	BRK2L
		13	3.93	BRK2L
		2	3.02	BRK2L
		5	1.91	BRK2L
		4	1.58	BRK2L
		8	1.56	BRK2L
		12	1.47	BRK2L
		3	1.34	BRK2L
		11	.67	BRK2L
		1	.50	BRK2L
			3.25	(AVERAGE INTENSITY)
BRK7L:+H ((1 10) 4 5))	1/13	3	3.46	
		MOLS NOT SHOWING THESE PROCESSES		1 2 4 5 6 7 8 9 10 11 12 13
BRK7L:−H ((1 10) (4 5))	9/13	6	4.49	
		8	4.48	
		1	4.46	

116

Table 7-1 (*Continued*)

REPORT OF ALL PROCESSES HAVING EVIDENCE for ANDROSTANE

31 SEPARATE PROCESS LABELS

FULL PROCESS LABEL	OCCURRENCE RATIO	(MOLS SHOWING THIS PROCESS) MOL ID	INTEN SITY	PARTIAL REDUNDANCIES
		7	4.41	
		5	4.33	
		2	4.17	
		9	3.51	
		3	3.34	
		12	3.07	
			4.03	(AVERAGE INTENSITY)
		MOLS NOT SHOWING THESE PROCESSES 4 10 11 13		
BRK17H:+H ((17 13) (15 14))	10/13	10	.92	BRK4H:+H
		8	.88	BRK4H:+H BRK6L:+H
		11	.67	
		13	.67	BRK6L:+H
		4	.67	
		3	.66	BRK4H:+H BRK6L:+H
		1	.56	BRK4H:+H BRK6L:+H
		6	.55	BRK4H:+H BRK6L:+H
		12	.54	BRK4H:+H BRK6L:+H
		7	.53	BRK4H:+H BRK6L:+H
			.67	(AVERAGE INTENSITY)
		MOLS NOT SHOWING THESE PROCESSES 2 5 9		
BRK17H ((17 13) (15 14))	1/13	10	.98	BRK4H
		MOLS NOT SHOWING THESE PROCESSES 1 2 3 4 5 6 7 8 9 11 12 13		
BRK17H:-H ((17 13) (15 14))	10/13	3	6.01	BRK4H:-H BRK6L:-H
		7	4.78	BRK4H:-H BRK6L:-H
		6	4.66	BRK4H:-H BRK6L:-H
		8	4.56	BRK4H:-H BRK6L:-H
		10	4.14	BRK4H:-H
		13	3.94	BRK6L:-H
		4	3.54	
		11	3.12	

Table 7-1 (*Continued*)

REPORT OF ALL PROCESSES HAVING EVIDENCE for ANDROSTANE
31 SEPARATE PROCESS LABELS

FULL PROCESS LABEL	OCCURRENCE RATIO	(MOLS SHOWING THIS PROCESS) MOL ID	INTEN SITY	PARTIAL REDUNDANCIES
		12	3.09	BRK4H:−H BRK6L:−H
		9	.55	
			3.84	(AVERAGE INTENSITY)
		MOLS NOT SHOWING THESE PROCESSES 1 2 5		

Notes:
1. BRK0 is the molecular ion.
2. BRK1L is loss of methyl. This is totally redundant with loss of methyl from another site in process BRK2L.
3. BRK7L is a "unique" explanation of the data points in the spectra of molecules where it occurs at all. That is, there are no other processes that explain these data points.
4. BRK17H (with gain or loss of one H) is rarely a unique explanation. In many spectra, the peak that can be explained by BRK17H:+H, BRK17H, or BRK17H:−H can also be explained by either of two other processes.

of the androstane skeleton. The task of RULEGEN is to generalize these descriptions so that a smaller set of more general subgraph descriptions accounts for a significant portion of the MS data. The goal is to achieve the appropriate amount of generalization without overgeneralizing, i.e., to find rules describing fragmentations in terms of subgraphs that are general enough to explain many data points at once and specific enough to avoid false predictions. At the opposite extreme from the specific process definitions produced by INTSUM is the single most general rule stating that "all bonds break and all allowed transfers occur." This zero-order-theory rule accounts for all possible breaks but obviously predicts many peaks that do not appear. The plausible rules lie somewhere between this single, overly general rule and the INTSUM-generated processes.

The RULEGEN program systematically and selectively searches the space of possible rules,[2] with strong guidance from the data as well as from the semantic model of mass spectrometry. For purposes of the search, a candidate rule is represented as a pair of constructs called a break environment (or B/E) and a template, which are explained below. A template represents an operation that, when applied to a B/E, transforms it into a general rule. A template thus defines a whole class of rules (because it is an abstract description of many rules, as will be seen).

---

[2] One organization of the program and requisite representation of the search space are described here. Another representation and organization are discussed in Mitchell (1978), in which the key idea is to keep a set of admissible versions of each rule and update the most specific and most general boundaries of this set on examination of each new training instance. It is still the subject of current research.

The search of the rule space is implemented by selectively generating more and more refined templates and applying them to B/E's that are determined directly by the INTSUM output. (This method is much more efficient than searching the rule space by applying all possible templates to all possible B/E's.) By the way the space of templates is generated, some rules will not be considered during generation. The search is *coarse* during RULEGEN, following the heuristic that rules that are close approximations to final rules can be found quicker and refined later.

A *break environment* (B/E) is a pair of associated elements: (1) A set of bonds in a molecule that may break, cleaving the molecule into two parts, and (2) a description of the molecule in question that is complete within the limits of the available vocabulary.[3]

The program first forms the *base B/E set*, i.e., the set of all break environments resulting from INTSUM's analysis of the data. The maximum possible cardinality of this set is the product of the number of processes and the number of molecules. In general it will be smaller since not all processes will occur in all molecules. In our example, the base B/E set contains at most 468 members (36 processes $\times$ 13 molecules).[4]

Associated with each B/E is the set of data points INTSUM has listed as potentially explained by the break that the B/E defines. The B/E's associated with positive evidence are the only ones that RULEGEN sees. RULEGEN's goal is to identify significant common features of break environments so that general rules may be composed to describe several specific B/E's. We expect to have far fewer explanatory rules than B/E's in every case.

As mentioned above, the program also needs the templates or rule abstractions for guidance in its search for rules. A *template* is a one-dimensional specification of significant attributes of a B/E whose values are to remain fixed in generalizing the B/E into a rule. For example, a template may specify that atom type is a significant determiner for fragmentation at a particular location.

In denoting templates, we will use "*" to denote a cleaving bond and "-" to denote a noncleaving bond, and we will adopt the convention that the charged fragment is on the left of a cleaving bond. If no particular features are specified for a node, it will be denoted by X. Otherwise, a node will be denoted by a sequence of from one to three letters separated by commas. The letter T indicates that the *atom type* of that node is significant. The letter N denotes that the number of (nonhydrogen) *neighbors* of that node is significant. The letter D denotes that the *number of dots* (double bonds) is significant for that node. For example, the template

(T1)   T,N - T,N * T

when applied to B/E

```
                         C
                         |
(BE-1)    C - O - C - O * C - C
```

[3] As indicated in earlier publications, we previously considered only B/E's of limited sizes, but this restriction has now been removed.

[4] See INTSUM's list of processes in Table 7-1. BRK0 is not included because it breaks no bonds.

generalizes it into the following rule:

$$\text{(R1)} \quad \begin{array}{c} X \\ \diagdown \\ \diagup \\ X \end{array} C - O * C$$

That is, the atom type of BE-1 has been fixed at three positions, and the number of neighbors has been fixed at the outermost position on the left. The size of the template also determines the size of the subgraph in the rule. In R1 all other features have been abstracted away in the generalization.

Templates are used to guide the exploration of the rule space within constraints of the positive evidence associated with B/E's. Only candidate rules for which there is some positive evidence are ever generated, because templates are only applied to the base B/E set. This limitation is a significant advantage over simple generate-and-test schemes in which rules are generated under the model and tested to see whether or not there is evidence supporting them.

In order to capture the notion of a candidate rule explaining more than one data point, we need to determine when two or more B/E's are equivalent under a template, or match with respect to a template. Two B/E's match with respect to a template if, within the size of the template there is a one-to-one correspondence between their atomic nodes such that (1) corresponding nodes have corresponding neighbors and (2) corresponding nodes have the same value for the features (T,N, or D) specified by the template for their positions. For example, both of the hypothetical B/E's below match under template (T1):

$$\text{(BE-1)} \quad \begin{array}{c} \quad \quad \quad \quad \quad C \\ \quad \quad \quad \quad \quad | \\ C - O - C - O * C - C \end{array}$$

$$\text{(BE-2)} \quad \begin{array}{c} \quad \quad C \quad \quad \quad \quad C \\ \quad \quad | \quad \quad \quad \quad | \\ C - C - C - C - O * C - O - C \end{array}$$

Applying (T1) to either or both (BE-1) or (BE-2) will result in rule (R1) above.

Thus, rules are explored by applying templates to B/E sets in order to determine the plausible rules that explain many data points. The most general template X*X is first applied to the base B/E set providing general rules that say, in effect, (1) every individual bond breaks (if the molecule is thereby cut into two fragments), (2) every pair of bonds breaks together (if the molecule is thereby cut), and so on. The next step is to refine this parent template systematically and determine which of the B/E's in the base B/E set are equivalent under each of the refinements.

A set of refined daughter templates is generated from a single parent template by adding specifications to the current template in accordance with the following rule:

Add exactly one of the attribute names T, N, or D to any current position or to a new position adjacent to a current outermost position with N specified, provided that the attribute name is not already specified at that position.

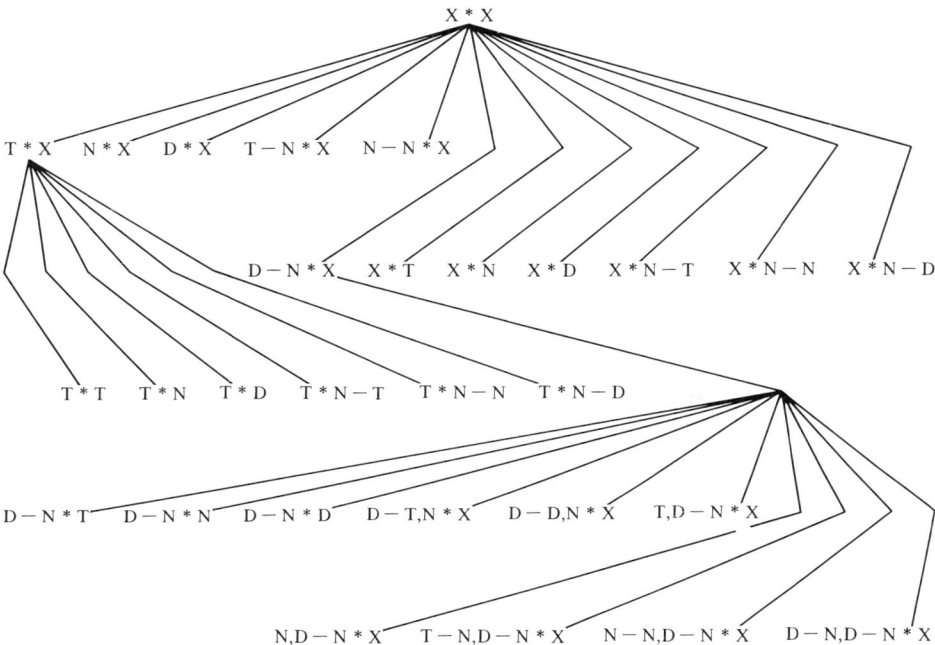

Figure 7-3 Template refinement with rulegen.

For example, X*X yields these 12 daughters: T*X, N*X, D*X, T-N*X, N-N*X, D-N*X, X*T, X*N, X*D, X*N-T, X*N-D, X*N-N. (See Figure 7-3.)

A template may *partition* a set of B/E's into more than one equivalence class if members of the set do not match with respect to the template. For compounds containing several carbons and one oxygen (such as monoketoandrostanes) the subgraph template T*T could partition a set of B/E's into at most 15 equivalence classes, if we limit consideration to single and double breaks. For one-break B/E's three nonequivalent environments are possible: C*C, C*O, and O*C (with O*O ruled out because there is just one O). For double break B/E's there are 12 nonequivalent environments containing at most one oxygen:[5]

```
     C*C        C*C        C*C        C*C
                 |                     | |
     C*C        C*C        C*C        C*C
    ----------------------------------------
     C*C        C*C        C*C        C*C
                 |                     | |
     C*O        C*O        C*O        C*O
    ----------------------------------------
     C*C        C*C        C*C        C*C
                 |                     | |
     O*C        O*C        O*C        O*C
```

[5] Note that the size and connectivity of the subgraph is fixed by the template, as well as atom type.

For more complex templates the possibilities escalate rapidly, even when there are strong structural constraints, as for ketoandrostanes. A particular B/E set may be subdivided into fewer than this maximum number of subsets depending on the actual structures, and in particular may not be subdivided at all.

As mentioned, each B/E set has an associated set of *evidence*, namely those peaks in the spectra that INTSUM could explain by the break (or process) in the B/E. It is summarized for RULEGEN as (1) the number of molecules associated with the set, (2) the average number of peaks per molecule, and (3) the maximum number of peaks per molecule.[6] These values will be used to compare B/E sets to direct the search toward plausible rules.

Beginning with the base B/E set and the most general template, X*X, the program *partitions* the base B/E set. Here the only relevant bases for partitioning are number of breaks and connection patterns of atoms within the template radius. The equivalence classes so produced are the first entries on a list of classes to be examined. Having selected a class for examination, every possible refinement of the present template is made to produce the next level of the template tree. At the first step this refinement will be T*X, D*X, N*X, etc. (as above). Each of these daughter templates may partition the selected B/E class, producing more specific candidate rules. The program considers each rule in turn and discards any duplicates. It also discards any rules that are not "improvements" over the parent rule according to the following definition. A rule is an improvement over its parent if three conditions hold:

1. The daughter rule predicts fewer ions per molecule than its parent (i.e., the daughter is more specific).
2. The daughter rule predicts fragmentations for at least half of all the molecules (i.e., it is sufficiently general).
3. The daughter rule predicts fragmentations for as many molecules as its parent, unless the parent rule was "too general" in the following sense: the parent predicts more than 2 ions in some single molecule or, on the average, it predicts more than 1.5 ions per molecule.

The remaining rules (B/E classes and templates) are added to the list of those to be explored further. When a rule chosen for examination (1) contains only one B/E, or (2) has only one peak or peak group per molecule in its evidence set, or (3) has an associated template that cannot be refined further, or (4) on refinement yields no new rules that are improvements [in the sense of conditions (1), (2), and (3) above], then it is considered a plausible rule and is saved for output.

Since each member of the B/E class has the same values of the attributes T, N, and D at each node for which its template specifies these features, the template instantiated with these values becomes the situation-part of the rule. The action-part is the B/E's definition of the break (plus transfers of hydrogen or neutral molecules) that accounts for the observed peaks in the positive evidence groups.

RULEGEN produced 12 rules for the monoketoandrostanes. The templates for

---

[6] In cases where a process includes transfers of atoms, a group of peaks can be associated with any single application of the process. In these cases, the program counts the average and maximum number of *peak groups* per molecule.

Table 7-2 RULEGEN templates for monoketoandrostanes

	Positive evidence			
Template	Number of molecules	Number of peaks	Average intensity	Maximum number of peaks
X*X	13	13	3.2	1
D–N*X	12	16	2.2	2
N–N*D	12	16	3.8	2
N*D	11	15	1.7	2
X*N	12	15	1.6	2
N*N–D	13	16	4.1	2
D*N	12	15	1.6	2
N*X	11	11	5.0	1
X*N–N	11	11	4.1	1
D–N*X	12	14	2.6	2
X*N–N	11	14	3.6	2
N*N	11	11	5.4	1

these rules are given in Table 7-2, along with the summary of the positive evidence for each rule. The rules themselves are not given since examples are shown at the end of the next section.

7.4 RULEMODification

The final phase of Meta-DENDRAL refines RULEGEN's set of rules to increase its generality and economy. This process is performed by RULEMOD, the testing phase of the plan-generate-test paradigm. Many of the refinements done here could have in principle been incorporated into the generating phase, but because of the combinatorial expansion, they are instead done only after a set of viable candidate rules has been generated. RULEMOD will typically turn out a set of 8 to 12 rules covering substantially the same data as an original set of 25 to 100 rules, but with fewer incorrect predictions.

Each rule generated by RULEGEN has been developed in isolation. RULEMOD attempts to improve and simplify the *set* of productions by eliminating redundant explanations. It also considers negative evidence, using it to further specialize overgeneral rules.[7] There are five steps to this procedure:

1. *Select* the most important rules.
2. Of those rules selected, *merge* similar rules.
3. *Specialize* rules of the resulting set to eliminate unconfirmed predictions.
4. *Generalize* these rules to increase their predictive range.
5. *Select* the most important rules from the modified set.

[7] Because of the cost of computing negative evidence for each candidate rule, this further process is not done in RULEGEN. In principle it could be.

7.4.1 Selection

Each rule is evaluated by computing a score according to the following equation:

$$\text{Score} = I \times (P + U - 2N)$$

where P = positive evidence: the number of times there is any evidence for the rule. (If any or all transfers associated with a rule produce peaks that are found in the spectrum, P is incremented by 1.)

N = negative evidence: the number of times a rule applies to a molecule whose spectrum contains no evidence for that fragmentation.

U = unique evidence: the number of times there is any *unique* evidence for a rule in the range of transfers. (Unique evidence is found if there are peaks accounted for by this rule and no others.)

I = (average) intensity of positive evidence: the ratio of intensities of peaks counted as positive evidence to P.

The rule with the best score is selected and the evidence supporting it is removed from further consideration so that it does not improve the score of other rules. The operational heuristic here is that no data points need two explanations, or, in other words, a rule should not gain merit by duplicating the explanations of other rules. The remaining rules are then reevaluated to select another rule. The process repeats until all scores fall below a threshold or until all remaining evidence is accounted for by one of the selected rules.

7.4.2 Merging

The selection procedure removes any rules that are totally redundant, but it is still possible that two (or more) rules explain many of the same peaks. If such rules are found, the merging procedure attempts to replace them with fewer, more general rules that account for the same data without introducing any new negative evidence (predicted peaks that are not found in the data). The modified rules are merged by searching for a common environment description that includes all the conditions found in the to-be-merged rules.

7.4.3 Specialization

The selected and merged set of rules is next modified to eliminate predicted evidence that does not occur in the data. This modification is done by making situation-parts more specific by adding one feature at a time. Any such refinement is kept if it deletes some unconfirmed predictions without loss of any positive evidence. This process is part of the *refined search* of the rule space that complements RULEGEN's *coarse search*.

7.4.4 Generalizing

One neighborhood-defining feature at a time is now removed from the situation-parts. If no new unconfirmed predictions are introduced by a deletion, the deletion is made permanent. This procedure may also add new predictions for which there is support in the data. This procedure is another important part of the refined search of rules in the local space of a rule found by RULEGEN, introduced to make rules more general.

7.4.5 Final Selection

The final step is a repeat of the initial selection step to remove new redundancies that have appeared in the process of merging, specializing, and generalizing.

A portion of the final set of rules for the monoketoandrostanes is given in Figure 8-4.

7.5 SUMMARY

Meta-DENDRAL is a learning program in the sense that it induces a set of general productions (describing the behavior of a class of compounds in the mass spectrometer) from specific instances. The program organization is a form of the plan-generate-test paradigm. As with Heuristic DENDRAL, there is a generator that must be constrained by planning and, for reasons of computational efficiency, is permitted to generate a set of first approximations that are then refined in the testing phase.

The initial set of specific instances delimits the class of hypotheses to be considered. Commonalities among the specifics are sought and used to generate a set of candidate-solution components (individual rules). Much computation is saved by limiting search to finding rough approximations rather than well-honed (and computationally expensive) final products, and by evaluating each component (production) rule in isolation. Only after the selection of a set of solution components that is very, very small in comparison to the potential space of possibilities are interactions among the components considered and detailed refinements of the solution made.

If we reexamine the motivations mentioned in the introduction to this chapter, we find the following observations are in order. First, Meta-DENDRAL certainly has succeeded in facilitating the application of DENDRAL to new compounds. The setup time for a new class of compounds is nominal, a few days in comparison to the weeks or months required by Heuristic DENDRAL. Second, new insights about the behavior of the molecules investigated (e.g., mono-, di-, and tri-ketoandrostanes) have indeed emerged; these details will not be reported here because they are technical and primarily of interest to chemists [see Buchanan et al. (1976) and also Chapter 8 of this book]. Third, our personal intuitions about the means by which scientists acquire and use specialized knowledge have certainly been deepened and enriched. But while the general organizational principles of Meta-DENDRAL have been demonstrated to work, it would be an exaggeration to claim that more than a scratch has been made on the surface of the complex problem of induction. It is, however, an interesting scratch that we feel merits deepening. We will have more to say on this issue in the final chapter.

CHAPTER
EIGHT
RESULTS

The results obtained by using DENDRAL are of primary interest to chemistry; the the design principles underlying the program are of primary interest to Artificial Intelligence. As an aid for systematic exploration of chemical structures, DENDRAL is unique. It has been applied to enough chemical problems to demonstrate its power and utility. In addition, the lessons learned from developing the system can be applied to the construction of other complex problem-solving systems.

8.1 INTRODUCTION

In this chapter we gather many of the results obtained to date using various versions and components of the DENDRAL system. Our purpose is to give readers some idea of the range and power of this system so that they may assess the results and potential of this research. The importance to chemistry of the particular results tabulated herein can best be assessed by those readers who are suitably knowledgeable about the appropriate aspects of organic chemistry and mass spectrometry. However, it should be possible, from an unhurried examination of this material, to gain an appreciation for the abilities of the DENDRAL system, present and future.

As has been emphasized, no one program is called DENDRAL. It should already be clear that DENDRAL is not a special purpose system for solving just those problems to be described shortly. Heuristic DENDRAL is a *framework* for helping chemists with structure elucidation problems in various ways. Some of the knowledge embodied in the system, such as the stability knowledge codified in the a priori GOODLIST and the a priori BADLIST, is general. This is also the case for the basic mass spectrometry theory that is embodied in the PLANNER and PREDICTOR programs. The class-specific and problem-specific chemical knowledge used by Heuristic DENDRAL is supplied by the chemist-user; the programs, however, are intelligent enough to understand such specifications and make use of them.

In addition to being an MS problem-solving engine of various configurations, the programs can be used in a more general vein as aids to chemists with structure elucidation problems. CONGEN, for example, is a symbolic graph manipulator for the chemist analogous to algebraic and analytic symbolic manipulators for the mathematician, such as REDUCE [Hearn (1971)] and MACSYMA [Martin and Fateman (1971)]. This chapter illustrates some of the applications in which the DENDRAL programs have aided chemists.

Finally, design lessons that have been learned from this work are discussed. These results are contributions to the art of heuristic programming and as such should be of greatest interest to the computer scientist.

8.2 THE SCOPE OF STRUCTURAL ISOMERISM

The concept of structural isomerism is basic in organic chemistry. The complexity of the combinatorics involved is such, however, that the variety and identity of isomers as a function of molecule size and other factors is still not fully understood. Rouvray (1974) summarizes various mathematical approaches to the question of numbers of isomers. Many lessons remain to be learned about the nature of the space of structural isomers and the effects on chemical variety of factors such as valence, ring structure, and heterogeneity of atomic types, among others.

The unconstrained DENDRAL generator provides means not only for determining empirically the number of isomers for a given empirical formula, but for identifying the structure of each. CONGEN provides in addition a means to determine the number of isomers with specific characteristics. For large problems these experiments are impractical. However, the cyclic and acyclic generators have been used to obtain some indication of the size of the spaces involved. The numbers turn out to be enormous, in fact often larger by an order of magnitude than was estimated by professional chemists.

The important trends are well illustrated in Figures 8-1 and 8-2, which are derived from CONGEN unconstrained runs. Figure 8-1 illustrates that the number of isomers increases approximately exponentially with number of carbon atoms, the rate of increase being larger with more rings and double bonds. Figure 8-2 plots the number of isomers as a function of unsaturations (i.e., number of rings plus double bonds) for compounds with a fixed total number of nonhydrogen atoms divided in various ways among carbon, nitrogen, and oxygen. The role of carbon in increasing the variety of compounds is well illustrated here. The smooth relation between number of isomers and degree of unsaturation is also interesting. Perhaps most striking is the extrapolation of the curves of Figure 8-1 to compounds containing larger numbers of carbon atoms, such as commonly concern chemists. The size of the space of possible structures quickly exceeds the practical limits for exhaustive search. This limitation is true for acyclic isomers alone also, as shown in Lederberg et al. (1969b).

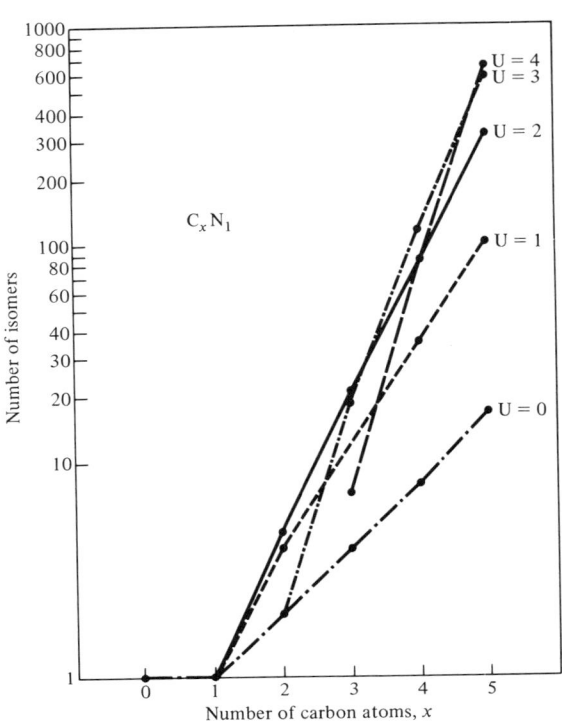

Figure 8-1 Semilog plot of number of isomers versus number of carbon atoms. (*Source:* Journal of Chemical Information and Computer Sciences, *Copyright 1975, American Chemical Society. Reproduced by permission. After Smith* [1975b].)

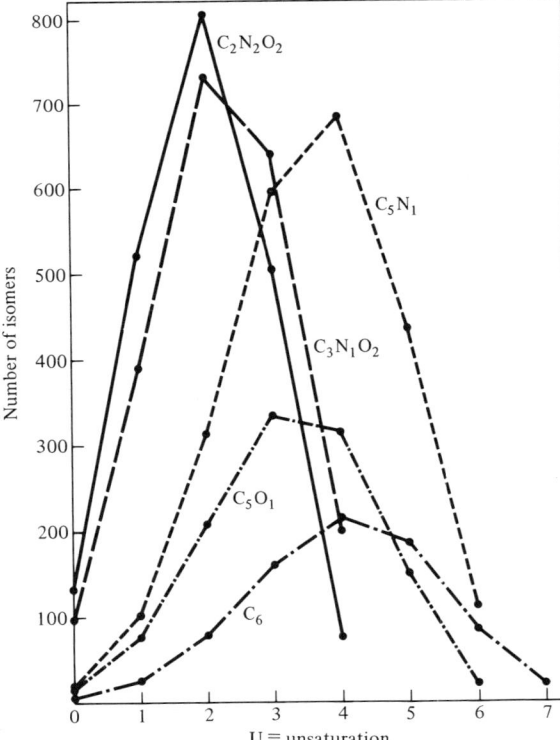

Figure 8-2 Number of isomers versus degree of unsaturation. (*Source:* Journal of Chemical Information and Computer Sciences, *Copyright 1975, American Chemical Society. Reproduced by permission. After Smith* [1975b].)

In addition to indicating the size of the problem space for structure elucidation, results have been generated to indicate the pruning power of additional chemical knowledge given to the DENDRAL programs. Table 8-1 lists the numbers of isomers for each of the 15 functional groups relevant to the empirical formula ($C_8H_{16}O_2$). The total number of isomers of this composition is in the thousands, so knowing which functional groups are present and/or absent will constrain the search immensely. BADLIST pruning constrains these numbers still more; the usual a priori BADLIST for

Table 8-1 Number of aliphatic isomers of $C_8H_{16}O_2$ selected by functional group

No.	Functional group name	No. of isomers of $C_8H_{16}O_2$	Contained subgraph(s)
1	Acid	39	—COOH
2	Ester	105	—COO—
3	Keto ether and aldehyde ether	329	≥COC≤ and —CO—
4	Hydroxy ketone and hydroxyaldehyde	458	≥COH and ≥CCO—
5	Diether (excluding enol ether)	183	(≥COC≤)$_2$
6	Hydroxy ether	783	≥COC≤ and ≥COH
7	Enol and ether	305	≥COC≤ and >C=COH
8	Hydroxy enol ether	497	≥COH and >C=COC≤
9	Unconjugated acetal	102	≥CC(OC<)(OC<)
10	Conjugated acetal	46	≥CC(OC<)(OC=C<)
11	Acyloin enol ether	48	≥COC=COC≤
12	gem-Diol	262	≥CC(OH)(OH)
13	Diol (excluding gem-diol and enol)	32	(≥CCOH)$_2$
14	Unconjugated peroxide	197	≥COOC≤
15	Unconjugated hydroperoxide	306	≥COOH

Source: Lederberg et al. (1969b). Reprinted with permission from the *Journal of the American Chemical Society*, copyright by the American Chemical Society.

acyclic structures is as follows (with "≡" denoting triple bonds):[1]

```
C=C-O-H        H-O-C-O-H
C=C-N-H        H-O-C-N-H
C≡C-O-H        H-C-N=O
C≡C-N-H        H-O-C=N

N=N-N          H-N-C-N-H
N=N-O          |
N=N-H          H

O-O            O=C-O-H
O-O-O          |
O-N-O          O
N-N-N          |
N-N-O
N-O-N          O=C-O-H
N-O-O          |
               O
               |
```

8.3 ACYCLIC HEURISTIC DENDRAL

Of the enormous number of classes of chemical compounds that might be grist for the DENDRAL mill, only a handful have passed through it. It may be of interest to know on what basis the selections were made.

The first class of compounds attacked was the amino acids. This class was basically a target of opportunity, for spectra had been collected from them for another purpose: the feasibility of a robot detector of life was being studied in connection with a preliminary design for a Mars probe. (The final design, as realized in the 1976 Viking mission, took a different approach to the life detection problem.) It might appear that the amino acids would prove to be a difficult class of compounds for the infant DENDRAL system because of their variety and the number of heteroatoms contained. Such proved not to be the case, because the absence of long, unbranched hydrocarbon chains in amino acids made them particularly appropriate for the primitive (zero-order) theory of mass spectrometry with which DENDRAL began. (The *zero-order theory of mass spectrometry* states that each bond will break, one at a time, so that all possible fragments will appear in the spectrum.) That this theory closely approximated the behavior of the amino acids was a happy state of affairs resulting in an early success, encouraging further work.

[1] It should be noted that while the a priori BADLIST for acyclic structures contains about 20 (forbidden) subgraphs, the chemistry of cyclic structures does not seem to permit strong a priori statements about unstable structures.

Professor Carl Djerassi, an organic chemist on the Stanford University faculty, was intrigued by the program's success with the amino acids but skeptical of its methods because its zero-order theory of mass spectrometry was so shallow. He proposed aliphatic ketones as an appropriate next test case, on the grounds that this class was in fact simple in structure and no appreciable challenge to a human mass spectrometrist, but would require significant extensions to DENDRAL's simplistic theory. This proved to be the case. Some of Djerassi's knowledge of ketones was captured in a revised version of DENDRAL, and ketones became the first in a series of classes of increasing complexity examined by the early versions of DENDRAL. Carl Djerassi became a highly valued colleague on the DENDRAL Project.

The sequence of classes examined in this early work proceeded in what is a fairly logical series of increasing complexity through various saturated, acyclic monofunctionals. After the ketones, the ethers and alcohols were chosen as the next most complex oxygen-bearing compounds. The oxygen of valence 2 constrains the possible structures more than a valence-3 atom, so the next most complex class to be selected was the aliphatic amines, which are acyclic nitrogen-containing compounds. Sulfur-containing compounds (thiols and thioethers) were examined to demonstrate the generality of the planning program.

Tables 8-2, 8-3, and 8-4 present results with amino acids, ketones, and amines, respectively. Results for ethers and alcohols, thioethers and thiols, are similar and are tabulated in Buchanan and Lederberg (1972). The reduction in the number of structures with planning is striking, particularly for the more complex compounds. Note

Table 8-2 Amino acid results without planning

Name of "unknown" amino acid	Chemical formula	Number of possible structures†	Number of plausible structures‡	Number of structures generated §	Rank order of correct candidate¶
Glycine	$C_2H_5NO_2$	38	12	8	1st, 7 excluded
Alanine	$C_3H_7NO_2$	216	50	3	1st
Serine	$C_3H_7NO_3$	324	40	10	1st, 9 excluded
Threonine	$C_4H_9NO_3$	1758	238	2	1st
Leucine	$C_6H_{13}NO_2$	10000 (approx.)	3275	288	Tied for 2d, 277 excluded

†The total number of possible structures is the number of topologically possible (and distinctive) molecular structures generated by the algorithm within valence considerations alone.

‡The number of plausible structures is the number of molecular structures in the total space that also meet the a priori conditions of chemical stability on BADLIST. The a priori rules have greater effect with increased numbers of noncarbon, nonhydrogen atoms.

§ The number of structures generated is the number of molecular structures actually generated by the program as candidate explanations of the experimental data. Pruning has been achieved by using the zero-order theory during structure generation.

¶ The rank order of the correct structure is the evaluation program's assignment of rank to the actual molecular structure used as a test "unknown." The number of structures excluded in the validation process is also indicated.

Source: Buchanan and Lederberg (1972). Reprinted by permission from *Information Processing '71*, copyright by Elsevier-North Holland Publishing Co.

Table 8-3 Ketone results with planning and testing

Name of "unknown" ketone	Chemical formula	Number of plausible structures†	Number of structures generated‡	Rank order of correct candidate§
2-Butanone	C_4H_8O	11	1	1st
3-Pentanone	$C_5H_{10}O$	33	1	1st
3-Hexanone	$C_6H_{12}O$	91	1	1st
2-Methyl-hexan-3-one	$C_7H_{14}O$	254	1	1st
3-Heptanone	$C_7H_{14}O$	254	2	Tied for 1st
3-Octanone	$C_8H_{16}O$	698	4	1st
4-Octanone	$C_8H_{16}O$	698	2	1st, 1 excluded
2,4-Dimethyl-hexan-3-one	$C_8H_{16}O$	698	4	Tied for 1st, 1 excluded
6-Methyl-heptan-3-one	$C_8H_{16}O$	698	4	1st
3-Nonanone	$C_9H_{18}O$	1936	7	1st
2-Methyl-octan-3-one	$C_9H_{18}O$	1936	4	1st
4-Nonanone	$C_9H_{18}O$	1936	4	1st

†The number of plausible structures is the number of molecular structures in the total space that also meet the a priori conditions of chemical stability on BADLIST. The a priori rules have no effect with formulas containing a single noncarbon, nonhydrogen atom. Thus, this column also represents the total number of possible structures.

‡The number of structures generated is the number of molecular structures actually generated by the program as candidate explanations of the experimental data. Pruning has been achieved by using the planning information from the planning program.

§The rank order of the correct structure is the evaluation program's assignment of rank to the actual molecular structure used as a test "unknown." The number of structures excluded in the process is also indicated.

Source: Buchanan and Lederberg (1972). Reprinted by permission from *Information Processing '71*, copyright by Elsevier-North Holland Publishing Co.

the further marked reduction when NMR data were manually used to determine the number of methyl radicals present in the compound. These results are indicative of the beneficial effects derived from the combination of multiple sources of information. It is particularly noteworthy that in some cases a *single* candidate is isolated from an initially very large set of possibilities. In these cases, the reduction is a consequence of knowing exactly the number of methyl groups in the compound, an observation that can be made from NMR but rarely from MS data alone. The planning and testing programs used to generate some of these results are the Preliminary Inference Maker and the PREDICTOR described above (Chapters 5 and 6). Additional PREDICTOR results are described below in Section 8.7.

After the generalization of the special heuristics for these classes was completed [and reported in Feigenbaum, Buchanan, and Lederberg (1971)], the project reached a turning point. The problems it had solved were all from small, acyclic molecules with a single functional group, and the data were all low-resolution mass spectra. It appeared that DENDRAL was capable of harder problems if provided with the additional information of high-resolution spectra.

At this stage the general cyclic generator had not been developed, so the means

Table 8-4 Amine results with planning but without testing

Name of "unknown" amine	Size Cn	Number of plausible structures†	Number of structures generated‡ MS	Number of structures generated‡ NMR	Name of "unknown" amine	Size Cn	Number of plausible structures†	Number of structures generated‡ MS	Number of structures generated‡ NMR
n-propyl	C3	4	1	1	N-methyl-di-*iso*-propyl	C7	89	15	3
iso-propyl		4	2	1	n-octyl	C8	211	39	1
n-butyl	C4	8	2	1	Ethyl-n-hexyl		211	24	1
iso-butyl		8	2	1	1-methylheptyl		211	34	1
Sec-butyl		8	4	2	2-ethylhexyl		211	39	9
Tert-butyl		8	3	1	1,1-dimethylhexyl		211	32	4
Di-ethyl		8	3	1	Di-n-butyl		211	24	1
N-methyl-n-propyl		8	4	1	Di-*sec*-butyl		211	33	8
Ethyl-n-propyl	C5	17	5	1	Di-*iso*-butyl		211	17	5
N-methyl-di-ethyl		17	4	1	Di-ethyl-n-butyl		211	17	3
n-pentyl		17	4	1	3-octyl		211	26	2
iso-pentyl		17	4	2	n-nonyl	C9	507	89	1
2-pentyl		17	2	1	N-methyl-di-n-butyl		507	13	1
3-pentyl		17	5	1	Tri-n-propyl		507	2	1
3-methyl-2-butyl		17	4	1	Di-n-pentyl	C10	1238	83	1
N-methyl-n-butyl		17	4	1	Di-*iso*-pentyl		1238	109	16
N-methyl-*sec*-butyl		17	3	1	N,N-dimethyl-2-ethylhexyl		1238	156	9
N-methyl-*iso*-butyl		17	4	1	n-undecyl	C11	3057	507	1
n-hexyl	C6	39	8	1	n-dodecyl	C12	7639	1238	1
Tri-ethyl		39	2	1	n-tetradecyl	C14	48865	10115	1
2-hexyl		39	8	1	Di-n-heptyl		48865	646	1
Di-n-propyl		39	8	1	N,N-dimethyl-n-dodecyl		48865	4952	1
Di-*iso*-propyl		39	8	1	Tri-n-pentyl	C15	124906	40	1
N-methyl-n-pentyl		39	8	1	Bis-2-ethylhexyl	C16	321988	2340	24
N-methyl-*iso*-pentyl		39	8	2	N,N-dimethyl-n-tetradecyl		321988	3895	1
Ethyl-n-butyl		39	6	1	Di-ethyl-n-dodecyl		321988	2476	1
N,N-dimethyl-n-butyl		39	10	1	n-heptadecyl	C17	830219	124906	1
n-heptyl	C7	89	17	1	N-methyl-bis-2-ethylhexyl		830219	2340	24
Ethyl-n-pentyl		89	16	1	n-octadecyl	C18	2156010	48865	1
n-butyl-*iso*-propyl		89	11	1	N-methyl-n-octyl-n-nonyl		2156010	15978	1
4-methyl-2-hexyl		89	16	4	N,N-dimethyl-n-octadecyl	C20	14715813	1284792	1

†The number of plausible structures is the number of molecular structures in the total space that also meet the a priori conditions of chemical stability on BADLIST. The a priori rules have no effect with formulas containing a single noncarbon, nonhydrogen atom. Thus, this column also represents the total number of possible structures.

‡The number of structures generated is the number of molecular structures actually generated by the program as candidate explanations of the experimental data. Pruning has been achieved by using the planning information from the planning program.

MS = Number of structures when only mass spectrometry is used in planning.

NMR = Number of structures when NMR data are used in planning to infer the number of methyl radicals.

Source: Buchanan and Lederberg (1972). Reprinted by permission from *Information Processing '71*, copyright by Elsevier-North Holland Publishing Co.

for a general attack on more complex cyclic molecules was not at hand.[2] As a result, a class of cyclic molecules of interest to Djerassi, the estrogenic steroids, was studied using special purpose heuristics and working from high-resolution spectra available from Djerassi's laboratory. An extension of the Planning Rule Generator (Section 5.2) led to the first successes on cyclic compounds. With the development of the cyclic generator, this class of compounds, plus other steroids already under study for other reasons, were the natural first choices to which the new procedures could be applied and generalized.

[2] A specialized generator of single-ringed structures had been written (circa 1970) as an extension of the acyclic generator. Although it was used for some studies [Sheikh et al. (1970)], the general utility of DENDRAL still awaited a generator of arbitrarily complex structures (now called CONGEN).

8.4 CONGEN RESULTS

Of all the DENDRAL programs, CONGEN is clearly the most generally useful and has been applied to current research problems by chemists around the world. To a large extent, CONGEN has moved from a research program to a service program. This is not to say that all the research problems have been dropped or solved, but that the current level of development already provides a valuable service and a production version of CONGEN is reliable enough to provide that service.

The range of problems to which CONGEN is applicable is great. To date perhaps a few dozen applications have in fact been made. Each of these is a relatively specialized problem whose importance is only clear in the context of a larger chemical issue.

Applications of CONGEN are no longer under close supervision by DENDRAL project members. This fact speaks for the maturity of the program but makes it difficult to enumerate problems to which it has been applied. As mentioned, it is a very general program. It works well on "medium size" problems in which it is combining 8 to 10 building block substructures, i.e., any combination of several atoms and superatoms. If the chemist has interpreted his data in terms of large superatoms (along with other constraints) then the program can manage structures with 50 or more atoms. The setup time for a CONGEN run depends on the number of constraints, but ranges from 2 to 20 minutes. Computation time ranges from a few CPU-seconds to many CPU-minutes

The final number of structures is strongly dependent on the chemist's ability to aggregate separate atoms into superatoms and his ability to infer other structural constraints from the available data. Because of the subtle, and powerful, interactions among constraints, it is difficult to estimate the final number of structures that will be produced. Nevertheless, a user can interrupt CONGEN at any time to request an estimate of the number of structures remaining to be generated. Although the estimates are rather crude, they are valuable for indicating to the chemist that the program needs more constraints, for example, if 200 structures have been generated and CONGEN estimates that it is only 5 percent finished.

Some of the problems have been studied for tutorial purposes and for comparison of the CONGEN approach to results achieved by conventional means. Many problems of this type have been solved [e.g., Carhart et al. (1975b), Smith (1975a), and Carhart et al. (1975c)]. Since the answers were known in advance, the ability of CONGEN to list the correct structure, sometimes with added possibilities, increases our confidence in the correctness of the program. Several similar examples of various degrees of complexity have also been worked out. See Cheer et al. (1976).

Much of the collaboration to date has been informal and has been undertaken with the dual expectation of providing useful service while leading to new CONGEN research areas. Some examples of problems are listed here to give a sense of the range of structures for which CONGEN has been of service.

1. *Chemical constituents of body fluids (e.g., organic acids, amino acids).* Generate structures within constraints derived from knowledge of chemical isolation procedures and human metabolic processes to identify compounds in the GC/MS traces of patients with suspected metabolic disorders of genetic origin.

2. *Terpenoids.* Generate structures within constraints (including four to six large superatoms) to identify C_{15} and C_{20} compounds isolated from natural sources. These were also tested to see if candidate structures obeyed the isoprene rule that restricts interconnections of 5-carbon superatoms [Smith and Carhart (1976), Cheer et al. (1976)].
3. *Marine sterols.* Determine the sterols (within numerous constraints) that could be metabolic precursors of known sterols in marine organisms in order to guide identification of unknowns.
4. *Insect secretions.* Confirm the structural possibilities in problems involving the identification of insect hormones and other insect defense secretions.
5. *Rearrangement products.* Obtain all structural possibilities under available constraints to help solve the structures of chemical and photochemical rearrangement products of unsaturated hydrocarbons (e.g., a tricyclic $C_{11}H_{12}$ hydrocarbon) and polycyclic molecules [Varkony, Carhart, and Smith (1976)].
6. *Ion structures.* CONGEN has also been applied to the problem of generating the structures of gaseous ions. This application has been done, for example, for the case of triethylamine, $(CH_3CH_2)_3N^{+-}$. See Smith, Konopelski, and Djerassi (1976) for an explanation of the underlying mass-spectrometry fragmentation mechanisms of this and related compounds.
7. *Pharmacologic agents.* CONGEN has been used to help solve the structures of several compounds displaying pharmacologic activity including some nitrogen heterocycles, pesticide conjugates, and metabolic products of microorganisms.

Another potential use for CONGEN is as a check on the results of analyses done by other means. Although chemists, unaided by CONGEN, have hypothesized structures that agree with all sources of information about a compound, it is seldom possible to be certain that equally plausible candidates have not been overlooked. In several cases (selected for their complexity) CONGEN has found alternatives to published hypotheses. For example, it was found that four structures in addition to the one presented in a published result met the stated constraints and were in fact chemically plausible. The routine use of CONGEN in this capacity could ensure a higher degree of reliability in the published literature.

8.5 PLANNER RESULTS

The next group of results illustrates the use of PLANNER with estrogens, using the fragmentation analysis and definitions described in Section 5.5. First, results for a single typical example are given. Figure 8-3 presents the low-resolution spectrum of 16alpha-hydroxyestrone. Table 8-5 shows the results of the break analysis for this compound, and Table 8-6 presents the results of planning, indicating the structures supported by the spectral data, and showing that all but one set of possibilities was eliminated by the PREDICTOR.

Table 8-7 briefly summarizes results obtained for 45 estrogens, using variations of the same techniques. This figure illustrates the program's feedback loop: constraints are relaxed with each pass until the program finds at least one structure that is consistent with the data. That is, the data are reinterpreted under weaker and weaker constraints if no consistent interpretation can be found under the stronger constraints. The procedure can be closely controlled by the user.

Table 8-8 presents some results on the difficult problem of analyzing *mixtures* of compounds. PLANNER's results are compared with published results obtained by conventional methods. The determination of the molecular ion in these examples was not done with MOLION, but with an earlier and less general method. These results are important because they illustrate the power of combining pieces of evidence for every fragmentation and every molecular ion in a much more thorough and systematic way than is possible by hand. They also indicate that some problems involving mixtures of similar compounds may be solved by the program without requiring the chemist to separate the individual components of the mixture.

Table 8-9 presents results from MOLION as applied to several compounds and demonstrates that the correct M^+ was among the top four candidates selected by the program in each case. Table 8-10 summarizes the performance of MOLION with 11 classes of compounds, using low-resolution spectra. When the molecular ion is present in the spectrum, MOLION always includes it among its top three choices, generally first. For compounds where the molecular ion is not present in the spectrum, such as amino acid derivatives and barbiturates, the correct candidate is still generally among the top five. There are some cases where prediction is not good (such as when $M^+ - 73$ is the highest mass); but overall the probability is greater than 0.95 that the molecular ion will be among the top five predictions, according to the empirical studies summarized here.

It is of interest how much of the chemist-user's time is required for imparting the necessary knowledge for analysis of a new class of compounds. Roughly, PLANNER

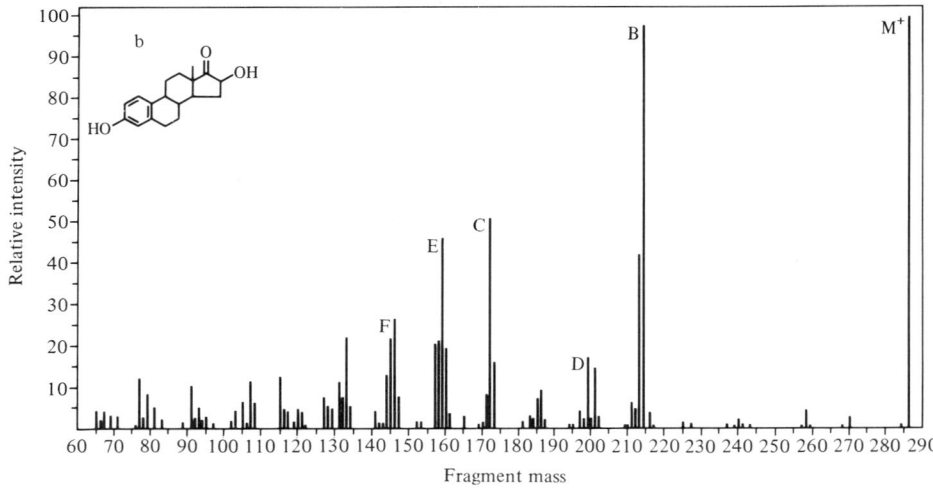

Figure 8-3 Low-resolution spectrum of 16alpha-hydroxyestrone. (*Source:* Journal of the American Chemical Society, *Copyright 1972, American Chemical Society. Reproduced by permission. After Smith et al. [1972].*)

Table 8-5 PLANNER break analysis for 16alpha-hydroxyestrone

Substituents	Break	Substituents on charged fragment	Summed relative abundance, %
03 DOT2	B	01	133
	C	01	60
	D	02	16
		01	15
		01 DOT2	5
	E	01	58
		01 DOT2	40
	F	01	45

Source: Smith et al. (1972). Reprinted with permission from the *Journal of the American Chemical Society*, copyright by the American Chemical Society.

Table 8-6 PLANNER results for 16-alphahydroxyestrone

Number of structures remaining after testing	Substituents	Placement
1	01	C-1–C-10
	01 DOT2	C-17
	01	C-16
2	01	C-1–C-10
	02 DOT2	C-17
3	01	C-1–C-10
	02	C-17
	DOT2	C-16

Source: Smith et al. (1972). Reprinted with permission from the *Journal of the American Chemical Society*, copyright by the American Chemical Society.

Table 8-7 Summary of results for estrogen standard compounds

Derivatives	No. of structures	One correct?	How obtained†
Of estrone (3-hydroxy-1,3,5(10)-estratrien-17-one)			
1, Estrone	1	Yes	Normal
2, 2-Hydroxy-	1	Yes	Normal
3, 2-Methoxy-	1	Yes	Normal
4, 3-Methoxy-	1	Yes	Normal
5, 1-Methyl-	1	Yes	Normal
6, 1-Methyl-3-methoxy-	1	Yes	Pass 2
7, 1,2-Dimethyl-	1	No	Normal
8, 1-Methyl-6α, 7α-dihydroxy-	2	No	Pass 2
9, 6-Methyl-	1	Yes	Pass 2

Table 8-7 *Continued*

Derivatives	No. of structures	One correct?	How obtained[†]
Of estrone (3-hydroxy-1,3,5(10)-estratrien-17-one)			
10, 7-Oxo-	2	Yes	Pass 2
11, 11-Oxo-9β-	5	Yes	Pass 3
12, 3-Methoxy-11α-hydroxy-	1	No	Pass 2
13, 15α-Hydroxy-	2	Yes	Pass 3
14, 16α-Hydroxy-	1	Yes	Normal
15, 6-Dehydro-	1	Yes	Normal
16, 7-Dehydro- (equilin)	1	Yes	Normal
17, 1-Methyl-6-dehydro-	1	Yes	Normal
18, 6-Methyl-6-dehydro-	1	Yes	Normal
19, 6-Dehydro-8-dehydro- (equilenin)	0		Pass 3
20a, 9,11-Dehydro- (mixture with 20b)	5	Yes	Pass 2
20b, Estrone	1	Yes	Normal
21a, 3-Methoxy-9,11-dehydro- (mixture with 21b)	4	Yes	Normal
21b, 3-Methoxy-	1	Yes	Pass 2
22, 1-Methyl-3-methoxy-9,11-dehydro-[‡]			
23, 17-Deoxo-	1	Yes	Pass 2
Of estradiol (1,3,5(10)-estratriene-3,17β-diol)			
24, Estradiol	1	Yes	Normal
25, 2-Hydroxy-	1	Yes	Normal
26, 2-Methoxy-	1	Yes	Normal
27, 3-Methoxy-	1	Yes	Normal
28, 1-Methyl-	1	Yes	Normal
29, 6-Oxo-	1	Yes	Normal
30, 11α-Hydroxy-	2	Yes	Normal
31, 3-Methoxy-11α-hydroxy-	1	No	Normal
32, 3-Methoxy-15α-hydroxy-	1	No	Pass 2
33, 16-Oxo-	3	Yes	Pass 3
34, 17α-Methyl-	1	Yes	Normal
35, "17α-Acetyl-"[§]	2		Pass 2
36, 1-Methyl-17α-acetyl-	3	Yes	Pass 2
37, 3-Methoxy-17α-vinyl-	1	Yes	Normal
38, 17α-Ethinyl-	1	Yes	Normal
39, 3-Methoxy-17α-ethinyl-	1	Yes	Normal
40, 1-Methyl-6-dehydro-	1	Yes	Normal
41, 1,2-Dimethyl-6-dehydro-	1	No	Normal
42, 9,11-Dehydro-	3	Yes	Normal
43, Estriol (1,3,5(10)-estratriene-3,16α,17β-triol)	1	Yes	Normal

[†]Normal: standard, one-pass processing. Pass 2: recycled with break B classification relaxed. Pass 3: recycled through entire program to include all evidence for all breaks.

[‡]See text for description.

[§]Compound is not estradiol 17α-acetate. The program indicates there is an extra unsaturation, possibly in ring C. The true identity of the sample is not known at this time.

Source: Smith et al. (1972). Reprinted with permission from the *Journal of the American Chemical Society*, copyright by the American Chemical Society.

Table 8-8 Comparison of PLANNER and conventional analysis for mixtures of estrogens

Mixture	Amount μg	Molecular ions	Estrogen planner	Conventional analysis
A	119	270 ($C_{18}H_{22}O_2$)		80% estrone
		300 ($C_{19}H_{24}O_3$)		20% 2-methoxyestrone
		286 ($C_{18}H_{22}O_3$)		Not reported
		298 ($C_{19}H_{22}O_3$)		Not reported
		284 ($C_{18}H_{20}O_3$)		Not reported
B	99	288 ($C_{18}H_{24}O_3$)		Estriol plus trace amounts of others
C	76	288 ($C_{18}H_{24}O_3$)		Estriol plus trace amounts of others
D	58	286 ($C_{18}H_{22}O_3$)		70% { 16α-hydroxyestrone, 16β-hydroxyestrone }
				25% 16-oxoestradiol-17β
				5% 15α-hydroxyestrone

Table 8-8 *Continued*

Mixture	Amount, μg	Molecular ions	Results	
			Estrogen planner	Conventional analysis
E	34	286 ($C_{18}H_{22}O_3$)	[structure: 16-hydroxy estrone type, 3-OH, 17-keto, 16-OH]	68% {16α-hydroxyestrone / 16β-hydroxyestrone} 23% 16-oxoestradiol-17β 9% 15α-hydroxyestrone
F	26	286 ($C_{19}H_{26}O_2$)	[structure: estradiol 3-methyl ether]	~70% {estradiol-17α 3-methyl ether / estradiol-17β 3-methyl ether}
		284 ($C_{19}H_{24}O_2$)	[structure: 11-dehydroestradiol 3-methyl ether]	~20% 11-dehydroestradiol-17α 3-methyl ether plus small amounts of several unknowns with one additional double bond
			[partial ring structures: ketone, alcohol, allylic alcohol fragments]	
		300 ($C_{19}H_{24}O_3$)	[structure: 11-oxo estradiol 3-methyl ether]	Unknown

140

Table 8-8 *Continued*

Mixture	Amount, μg	Molecular ions	Results — Estrogen planner	Conventional analysis
			[structure]	
			[structure]	
			[structure]	
G	24	270 (C$_{19}$H$_{22}$O$_2$)	[estrone structure]	~90% estrone
		300 (C$_{19}$H$_{24}$O$_3$)	[2-methoxyestrone structure]	~10% 2-methoxyestrone
		286 (C$_{18}$H$_{22}$O$_3$)		Not reported
H	14	286 (C$_{19}$H$_{26}$O$_2$)	[estradiol 3-methyl ether structure]	~80% { estradiol-17α 3-methyl ether; estradiol-17β 3-methyl ether }
		284 (C$_{19}$H$_{24}$O$_2$)	[structure with OH;U]	~20% 11-dehydroestradiol-17α 3-methyl ether plus several unknowns with one or two additional double bonds

Source: Smith et al. (1973a). Reprinted with permission from the *Journal of the American Chemical Society*, copyright by the American Chemical Society.

Table 8-9 MOLION predictions and rankings for illustrative compounds

Compound	Mol formula	Highest mass present	Fragment missing†	Mol wt	Ranked at no. ‡
Ritalin (2)	$C_{14}H_{19}NO_2$	172 (M − 61)	$C_2H_5O_2$	233	4
Pentobarbital (3)	$C_{11}H_{18}N_2O_4$	197 (M − 29)	C_2H_5	226	2
Mebutamate (4)	$C_{10}H_{20}N_2O_4$	175 (M − 57)	C_4H_9	232	3
Tridecan-7-one	$C_{13}H_{26}O$	155 (M − 43)	C_3H_7	198	4
Succinic acid methyl ester	$C_6H_{10}O_4$	116 (M − 30)	CH_2O	146	2
Caprylic acid methyl ester	$C_9H_{18}O_2$	129 (M − 29)	C_2H_5	158	3
Glutaric acid methyl ester	$C_7H_{12}O_4$	129 (M − 31)	CH_3O	160	1
Maleic acid butyl ester	$C_{12}H_{20}O_4$	173 (M − 55)	C_4H_7	228	2
N-TFA α-alanine § butyl ester	$C_9H_{14}NO_3F_3$	186 (M − 55)	C_4H_7	241	2
N-TFA norleucine butyl ester	$C_{12}H_{20}NO_3F_3$	227 (M − 56)	C_4H_8	283	2
N-TFA valine butyl ester	$C_{11}H_{18}NO_3F_3$	227 (M − 42)	C_3H_6	269	2
N-TFA threonine butyl ester	$C_{12}H_{15}NO_5F_6$	323 (M − 44)	C_3H_8	367	1
N-TFA phenylalanine butyl ester	$C_{15}H_{18}NO_3F_3$	216 (M − 101)	$C_5H_9O_2$	317	4
n-Undecyl alcohol	$C_{11}H_{24}O$	154 (M − 18)	H_2O	172	1
4-Methyloctan-4-ol	$C_9H_{20}O$	129 (M − 15)	CH_3	144	1

†These fragment composition losses from the molecular ion are only postulated. Their validity could only be confirmed by high-resolution studies.
‡Note "ranked at number 1" is the program's best choice for a molecular ion candidate.
§TFA refers to the trifluoroacetyl derivative.
Source: Dromey et al. (1975). Reprinted with permission from the *Journal of Organic Chemistry*, copyright by the American Chemical Society.

rules for a new class of simple compounds can be entered into the program in a few minutes, although it may take one or two weeks for a chemist familiar with the class and its MS behavior to decide on an appropriate set of rules. For the more complex cyclic classes such as the estrogens, the time requirements jump to one or two months, if program modifications are needed to accommodate new problems.

Table 8-10 Summary of MOLION performance

Class	M⁺ in top 3	M⁺ in top 5	Total number of compounds
Amines	62	67	68
Alcohols	51	54	57
Ketones	42	44	44
Ethers	33	34	34
Acetals	13	14	14
Amino acid derivatives	9	11	13
Thioethers	11	12	12
Drug compounds	6	8	9
Methyl esters	6	8	8
Butyl esters	4	5	6

Note: The program failed to generate the correct candidate for one alcohol, tert-butyl alcohol.

Source: Dromey et al. (1975). Reprinted with permission from the *Journal of the American Chemical Society*, copyright by the American Chemical Society.

8.6 META-DENDRAL RESULTS

The Meta-DENDRAL program has been used to study the MS behavior of several classes of compounds. Before it could be used with confidence on new cases, however, it was necessary to reproduce previous results on at least two widely differing classes of molecules. Low-resolution spectra for 11 aliphatic amines and high-resolution spectra for 10 estrogenic steroids were the two test cases chosen. Setup time for Meta-DENDRAL, once spectra have been collected, is less than an hour. However, the interpretation of results and refining of system parameters can take some weeks.

Meta-DENDRAL was applied to the low-resolution spectra of 11 *aliphatic amines* ranging in size from 4 to 14 carbons. Five rules were produced, explaining 84 percent of the total ion current. The five rules described processes that had previously been known to describe the behavior of these compounds in the mass spectrometer. They were (1) alpha cleavage, (2) beta cleavage with hydrogen transfer, (3,4) two cases of gamma cleavage with concomitant C—N cleavage, and (5) two-bond cleavage yielding a nitrogen-containing fragment.

Applied to the high-resolution spectra of 10 *estrogenic steroids*, Meta-DENDRAL produced eight rules that well characterize these compounds, accounting for over 40 percent of the total ion current. Five of these rules were the same as those found in the literature and used earlier for PLANNER (although Meta-DENDRAL was not in any way primed with these rules). The three additional rules describe cleavages through rings B and C of the estrogen skeleton (Figure 5-4, Section 5.5) that are plausible processes.

Meta-DENDRAL has also been successfully applied to three classes of *ketoandrostanes*. No fragmentation processes for these compounds taken as a class had previously been described in the literature. For monoketoandrostanes, eight rules were produced, accounting for 42 percent of the total ion current (74 percent of the data explained by

INTSUM). (We have used this case to illustrate the organization of the program; see Chapter 7.) For diketoandrostanes eight rules were also produced, accounting for 33 percent of the total ion current (77 percent of the data explained by INTSUM). The data were high-resolution spectra of 9 of the 55 possible diketoandrostanes. Ten rules were produced from the high-resolution spectra from 8 of the 165 possible *triketoandrostanes*; these rules accounted for 60 percent of the total ion current (84 percent of the data explained by INTSUM).

The rules for these classes of compounds are depicted in Buchanan et al. (1976). A few are shown in Table 8-11 for illustration.

8.7 DENDRAL PREDICTOR RESULTS

The prediction and ranking programs have been tested successfully on several sets of known compounds and used to aid in structure elucidation of some unknowns, as mentioned in Section 6.4, including monoketoandrostanes, sterols, and macrolide antibiotics. If we assume that the correct structure was in the CONGEN list, the ranking program was expected to pick it out among all plausible isomers by comparing predicted mass spectral peaks for each isomer with the observed mass spectrum (of the one correct structure). The predictor itself is a deductive mechanism that provides testable results, either by the half-order theory or by more complex, class-specific rules. The ranking function is a heuristic program that compares predicted and observed spectra to discriminate the structure whose predictions best match the observed data from all others.

Several scoring functions have been tried on test sets to determine the sensitivity of the ranking to the specifics of the scoring. All are variations on the simple idea that the score for a structure should be increased when predictions are confirmed (in the observed spectrum) and should be decreased when predictions are disconfirmed. Some functions give extra weight to predicted high mass peaks on the principle that these are more significant. Other functions give extra weight to disconfirmed predictions on the principle that negative evidence is more significant, since spectral peaks arising from *any* source can be used to confirm predictions. The net result of our experimentation with scoring functions is that simple functions do about as well as complex ones, especially if there are many predictions included in the score. The score is much more sensitive to the quality of the rules used for making predictions than it is to the composition of the scoring function. The half-order theory provides a reasonably efficient filter because it almost never makes numerous erroneous predictions on the correct structure, and it is specific enough to make predictions for the incorrect structures that will be disconfirmed. On the other hand, it is general enough that many candidates pass through this filter.

For the results mentioned below, the score for each comparison was computed as follows.

For a given set of rules, a particular structure-to-spectrum comparison yields a numerical *score*. This score is a function of

$S(R)$ = significance of a peak correctly predicted by rule R
 = mass of the corresponding ion $\div M^+$

Table 8-11 Meta-DENDRAL produced rules for monoketoandrostanes

Name†	Subgraph‡	Other descriptors and interpretations	Score§	Positive evidence¶ Any	Positive evidence¶ Unique	Negative evidence	Average intensity $\%\Sigma_{40}$
M-1 (none, +H, −H)		Atom a is not keto-substituted	145.8	24	11	1	4.42
M-2 (none, +H, −H, −2 H)			135.3	21	12	3	5.01
M-3 (none)		Loss of methyl	84.5	26	26	0	1.62
M-4 (none, −H, −2 H)			70.5	21	0	1	3.71
M-5 (+H, −H)			39.5	8	4	0	3.29
M-6 (−H)		There must be a keto group on atom a	39.7	9	0	0	4.41
M-7 (−H, −2 H)		Atoms a and d are not keto-substituted	23.8	13	13	7	1.99
M-8 (+H, −H)		Atom d is not keto-substituted	13.1	3	0	0	4.35

†Important transfers of neutral species are indicated below the name of the rule. Absence of transfers is stated explicitly as "none."
‡Specified nonhydrogen substituents are indicated by R (R ≠ H). Other valence positions may be filled with any atoms (including H) except when restricted by other descriptors.
§Score calculated as described in Section 7.4.1.
¶Positive evidence count (any and unique) and negative evidence count are described in Section 7.4. The number of positive instances may be greater than the number of molecules because a rule may apply more than once in any molecule.
Source: Buchanan et al. (1976). Reprinted with permission from the *Journal of the American Chemical Society*, copyright by the American Chemical Society.

and

> $K(R)$ = number of peaks in the spectrum that are correctly predicted by application of rule R to the structure, or -1 if there are no correctly predicted peaks.

$K(R)$ will be greater than 1 if the rule predicts lower m/e peaks due to transfers, and these are indeed found in the spectrum along with the major peak predicted.

The *score* is the sum of $[S(R) \times K(R)]$ over all the rules in the set. The scoring function used in discrimination penalizes a candidate structure if its predicted spectrum shows significant peaks that are not in the actual data. The other kind of mismatch—failure to predict peaks that appear in the actual spectrum—does not penalize a candidate. The reason for this asymmetry is that the rules are selected for their generality and thus should make correct predictions, but they are not expected to explain all ions in a spectrum. When a predicted mass spectrum from a candidate structure is compared with an actual spectrum, then the peaks that arise from non-general fragmentation processes will not be predicted.

Using the eight estrogen rules and this scoring function, we showed that the program can distinguish several of the estrogenic steroids from all other isomers of the same class.

For the ketoandrostanes, this scoring function was used to test the discriminatory power of the respective rules by testing their ability to rank known structures higher than other possibilities. Rules M-1 through M-8 (six of which are shown in Figure 8-4) have relatively low discriminatory power for the monoketoandrostanes. These rules can distinguish the 7- and 11- keto compounds from among all other possible monoketones, but they ranked the other structures anywhere from second to last when comparing the spectrum of the correct structure with predicted spectra for all candidates. This low discriminatory power relative to the estrogens is due to a combination of three factors: (1) the rules seldom mention the carbonyl group explicitly (by itself this is not necessarily bad), (2) each rule often predicts the same ions in all the molecules, and (3) where different ions are predicted they are not unique to the spectrum of the correct compound.

It is interesting that the eight diketoandrostane rules have much better discriminatory power than rules M-1 through M-8 had for the monoketoandrostanes. We compared the predicted spectra for the 55 possible diketoandrostanes (excluding substitution on the C-18 and C-19 methyl groups) against the actual spectrum for each of nine diketones. Three compounds were ranked first (i.e., discriminated correctly). The rest were ranked 6th, 3rd, 4th, 16th, 8th, and 9th respectively, out of 55 candidates. The improvement in discriminatory power over the monoketoandrostanes reflects more frequent (implicit) reference to keto groups in the rules and a larger number of unique ions predicted by the rules.

The discriminatory power of the 10 triketoandrostane rules is quite high, paralleling the discriminatory power of the diketoandrostane rules. The predicted spectra of the 165 possible triketoandrostanes were compared to the known spectrum of each of eight compounds. The ranking of the correct structure was 4, 5, 3, 4, 9, 20, 4, and 12, respectively.

These programs have been used to infer structures for sterols isolated from marine sources. CONGEN's lists of possibilities (within constraints specified by the chemist) contain several dozen structures. Ranking these with respect to a few known fragmentation rules has been an important aid to the chemists.

8.8 DESIGN PRINCIPLES

In *The Sciences of the Artificial* Herbert Simon (1969) distinguishes natural sciences from "artificial" sciences, which he judiciously calls sciences of the artificial. By the latter he means sciences of design—the systematization of knowledge that man employs in constructing artifacts. Although design is more often conceived as an art, Simon argues that an instructive comparison may be made between design principles on the one hand, and empirical generalizations about natural phenomena on the other. He suggests that the art of design could become the science of design. Since it has frequently been observed that the pursuit of scientific knowledge is itself an art, Simon's suggestion includes the case of the science of doing science.

The discoveries of natural science are facts and theories. The discoveries of the sciences of design are design principles. The exact status of design principles has never been analyzed, so there is something more compelling, more real, about a fact or empirical law than about a principle of design. Nonetheless design principles are of preeminent importance in human endeavor. Facts are seldom ends in themselves; more often their importance is in the service of design.

Viewed on the model of natural science, artificial intelligence may be properly criticized when it does not result in facts about computing, facts about cognition, or artifacts in the form of successful performance programs. But such results are only part of what should be expected from a science of design. It is because of the unclear status of design principles, we feel, that empirical studies of design, such as AI projects, have been subjected to harsh criticism for failure to produce substantive results. It is more than this unclear status, of course. Since we lack any systematic way of stating design principles and their range of application, these principles bear a certain will-o'-the-wisp quality that flits between the obvious and the false. There simply is no compelling way to convey just what the builder has learned; in the absence of a program of demonstrable power, as is more often than not the case, the suspicion is that nothing has been learned.

This book is about the discovery, use, and importance of heuristics. No theorems have been proved, no systematic experiments run, no general empirical laws recorded, but we have done a substantial amount of experimentation in the design of heuristic programs. Since a program of demonstrable power has been produced to do a job that scientists consider significant, interest attaches to the underlying principles of design; we know, at least, that they do work in a nontrivial case. In this section we will attempt to convey as clearly as possible, given the presystematic state of the science of artificial intelligence, some of the lessons that have been gained from this experience.

8.8.1 Considerations for Program Design

DENDRAL is an extensive case study of a knowledge-based program within the plan-generate-test paradigm. The success of DENDRAL illustrates the power of this paradigm. It also clearly points to a number of details of this methodology that deserve careful consideration in its application.

1. The efficiency of the generator is extremely important. Even with effective planning and testing, the power of the problem solver will often be limited by how quickly candidate solutions can be enumerated. It is particularly important that constraints can be applied effectively.

One interesting improvement in CONGEN involves spending a little extra time at the beginning of a session to save enormous amounts of time later. There are numerous, logically equivalent ways of specifying constraints to CONGEN that differ greatly in efficiency. It is unreasonable to expect a chemist using CONGEN to be familiar enough with the program itself to know which specifications are more efficient than others. (It is also unreasonable to ask the chemist to seek advice every time from one of the local CONGEN experts.) Thus we have begun work on a "smart interpreter" of constraints. The goal is to have it accept the chemist's statement of the problem and understand both the chemistry and the computational procedure well enough to transform the given constraints into an equivalent, more efficient set of specifications.

2. The use of depth-first search, which provides a stream of candidates, is generally better (in an interactive program) than breadth-first search, in which no candidates emerge for examination until all are generated. To a programmer the two methods are equivalent. To a chemist waiting for help with a structure problem the difference is substantial. It is far more pleasing to see some answers quickly, and it is more efficient in those cases where the first few answers reveal mistakes in the problem specification. In those cases, the chemist can interrupt the program, change the constraints, and restart.

3. Planning is in general not simply a nice additional feature but is essential for the solution of difficult problems. As much knowledge as possible should be brought to bear at this stage rather than at the testing stage, because this point is where the search can be cut drastically.

4. Every effort to make the program uniform and flexible will be rewarded. The user should be provided with as many options as one can think of (with defaults established to remove the burden when the flexibility is unneeded). Every decision strategy and parameter that is hardened into the program will become a limitation not open to examination or easy modification and not easily remembered.

5. Interactive user interfacing is not merely a nicety but an essential. For a high-performance computer program to capture the sustained, widespread attention of working scientists, it must contain a large number of features that make it easy and pleasant to use. These features are commonly termed "human engineering aspects" of a program. In very rare instances, a program will be so useful that it will be widely adopted even without proper attention to human engineering. More often, programs that are understandable only to programmers are used, if at all, only by programmers.

The prompts and descriptions printed by the programs have been designed by chemists to be terse, informative, objective, and courteous. These are not always consistent goals, but with careful attention the dialogue can be free of flagrant affronts to our feelings. Along the same lines, it should be noted that a person's right to privacy cannot be ignored in scientific programs. An automated laboratory notebook, for example, should be confidential if the chemist wishes it to be. In other contexts much work has been done on this problem. In CONGEN we rely on the standard protection mechanisms built into the computer system, but we will need more security if we are to satisfy users in chemical industries. As mentioned above, the system tries to be responsive to a user's request for help, and users are encouraged to call one of several local chemists with knowledge of the program in case of problems. The importance of talking to someone who understands the task is obvious to users of complex programs, but not always obvious to programmers.

6. An interesting extension of the plan-generate-test paradigm could improve its power: search and generation (see Section 3.2.2) might be combined into a single problem solver. In the context of DENDRAL this combination would mean that the generation of isomers would be guided by a search through a related problem space. The problem states would most naturally be chemical structure graphs, and the transformations would append, delete, and rearrange constituents. A proposed alteration would be considered for its effects on the mass spectrum, and a hill-climbing technique, for example, might drive the search. Successful ("warming") modifications would become GOODLIST constraints on generation. Numerous variations on this theme can be envisioned, and work has been started on one of them.

7. Choice of programming language is still an issue. We have yet to see a language that combines the flexibility and debugging power of INTERLISP with the running speed and exportability of FORTRAN. This language conflict causes a dilemma at the start of a large programming effort whenever the designers hope for widespread use of the resulting program. Networking provides a partial answer to the exportability question, since widespread use can be accomplished by long-distance sharing of a complex program.

8. Providing assistance to problem solvers is a more realistic goal than doing their jobs for them. In the first place it removes some of the psychological barriers that people often exhibit toward machines. Also, the amount of work involved in automating the whole task may far outweigh the benefits and in any case will delay the appearance of any benefits considerably. This is the theme of much of Norbert Wiener's writing [e.g., Wiener (1964)].

9. Record keeping is an important adjunct to problem solving. Every laboratory assistant is expected to keep a good laboratory notebook: the same should be true for a computer apprentice. Of the many ways of realizing the goal of helpful records, only some have been explored in the context of DENDRAL.

We expect the DENDRAL programs to provide three different kinds of notes: (1) A record of initial conditions, intermediate conclusions, and final results; (2) a complete record of the interaction between chemist and program (including false starts and typing mistakes); (3) a trace of the program's reasoning steps.

Each of these is important for a different reason. The final results, of course, are the sine qua non of the assistant's work. The record of initial conditions and major intermediate conclusions give the chemist at a glance the context in which the problem was solved and the major steps in its solution. This record serves as a useful reminder of scope and limits; in addition, disagreement on initial conditions or intermediate conclusions would be sufficient reason to request the assistant to start over. Meta-DENDRAL, for instance, immediately precedes its stored and printed results with a summary of the context specified by the chemist. Because they are together, there is less chance that the results will be interpreted without proper regard for the context.

The record of the chemist's interaction with the program is a detailed account of what the investigator requests of the assistant. Failure to find a solution to a problem can often be attributed to ill-specified requests, so it is helpful to review the complete record of specifications made by the investigator. For example, the requests for help and the program's responses are important entries in the experimental record.

Finally, the trace of the assistant's reasoning steps is helpful for keeping track of the inferential steps of an assistant that might otherwise not be open to scrutiny by the investigator. In any case it is often useful to have a record to justify moving from one point to the next. For example, before the DENDRAL PLANNER prints the final results, it prints the plausible molecular ions it inferred from the data and the data it associates with each of the separate fragmentations.

10. In order to use a program intelligently, a user needs to understand the program's scope and limits. The scope, roughly, is the broad class of problems that the program is designed to solve and the context in which solutions will be found. The limitations of a program are the idiosyncrasies that must be remembered to obtain reliable solutions and are less fundamental to the whole procedure. For example, enumerating polymeric structures is outside the scope of CONGEN, while its working definition of aromaticity is a limitation that is more easily changed. Operationally, the scope is the broad definition of the problem that can only be changed at the cost of writing an entirely new procedure. The limitations are the explicit and implicit items in the problem definition that are added to make the problem solvable and that may be changed or removed more readily. It is not a sharp distinction; the point is that a chemist needs to understand the program's interpretation of the problem before the program can be used responsibly and confidently.

11. The context in which problem solving proceeds is essential information for interpreting the solutions. The more an assistant can make explicit the assumptions and initial conditions of a problem, the easier it is for an investigator to understand the answers. This has always been true, but the emergence of computer programs as assistants brings the problem clearly into focus.

In a program the assumptions are often completely hidden. One effect of this fact has been to divide the scientific community into roughly three camps: scientists who mistrust all computer programs, scientists who will believe anything generated by a computer, and scientists who write their own programs. As computer science matures, however, this division will fade because programs will be able to convey many of their own assumptions and limitations.

The only step we have made along these lines with DENDRAL programs is to keep a good laboratory notebook, as described above. One of the items we try to make explicit at the time problem solutions are printed is the set of assumptions under which the program arrived at those solutions.

12. DENDRAL employs uniformity of representation as a means of understanding (and conveying) the contents of the knowledge base as well as problems of acquir-

ing new knowledge. In DENDRAL, knowledge is uniformly represented in a very general form: production rules. The uses to which the knowledge is put, however, are many. This arrangement achieves the best of both worlds: we have uniformity of representation with its virtues of modularity, simplicity of control structure, and perspicuity, and we have the inherent power of multiple sources adding to and making varied uses of the common knowledge base.

8.8.2 Applying the Lessons Elsewhere

Particularly in the physical sciences there is a great emphasis on precision and instrumentation. Mass spectrometry is no exception. A premium is placed on achieving greater and greater resolution of the instrument, on the elimination of noise and impurities, and on obtaining the cleanest possible data. Then and only then does one get on with the business of analysis.

Experience with DENDRAL suggests that there is, at least, an alternative approach. The same amount of effort spent on refining the means of hypothesis generation and testing will often yield far greater dividends. The power of hypothesize-and-test is unmatched by the method of construction (by deduction from previously verified premises), not only at the grand level of theory construction, but at the mundane level of day-to-day experimentation.

Constructive, deductive methods, in which each step follows from the last, have an understandable appeal. Such methods are conservative: they may fail but they will not err. This reason, we believe, operates against the acceptance of heuristic programming methods. If there is such a thing as a classical view of programming, it is that there can be no program without an algorithm. With the important exception of the algorithm underlying the cyclic structure generator, the programs we have described contain no glimmer of elegant, formal theory. MOLION, for example, would not pass muster as a theory, and yet it works. The most important and powerful knowledge any working scientist has is unformalized intuition about his field. Heuristic programming offers a means of capturing and amplifying this knowledge. Exploitation of this technique could put computers in the service of science to an extent that would eclipse the contributions of the traditional programming approaches.

A conscious effort was made to incorporate many of DENDRAL's design principles in another knowledge-based program written at Stanford, the MYCIN program [Shortliffe (1976)]. MYCIN incorporates knowledge of bacterial infections and antibiotic drugs in production rules to aid physicians with the task of selecting appropriate antibiotic therapy for patients. MYCIN's success is due, in part, to experience gained in organizing and manipulating large knowledge bases for DENDRAL.

In selecting research projects, the following guidelines have proved valuable: (1) Do not count on breakthroughs; (2) find a real star as your expert; (3) make sure your star will devote at least one-third time to the project; and (4) pick one who knows or is willing to learn about computers.

8.8.3 Project Organization

There have been few successful interdisciplinary projects in the history of science, but we believe DENDRAL should be counted among them. The project has worked cohesively for a decade, and it has involved in productive interaction researchers from the disciplines of chemistry, computer science, genetics, philosophy, physics, mathematics, electrical engineering, management science, and psychology.

It is difficult to give a recipe for this success, but we believe we can list some important ingredients. First, the task was conceived in such a way as to appeal to many interests; it could have been described as a "pure" mass spectrometry problem, or a "content-free" hypothesis formation problem, but it was not. This task is not prohibitively difficult: it can be understood (with a moderate effort) by anyone with a modest technical background. One scientist, with knowledge of both chemistry and computer science, was willing to coordinate and arbitrate the often-conflicting efforts of the group, and was able to do it because others felt sufficient respect for this man's ideas and vision to sacrifice some of the traditional autonomy and rugged individualism of scientists. Not the least important, a natural selection has occurred, resulting in a staff of specialists each of whom is truly willing to go more than half way to understand the other's discipline, paradigms, and arcane jargon.

We may offer no magic advice here, but the lessons are important, and mistakes are costly. In spite of the advantages this project has had, the going has been rough. Interdisciplinary work is antithetical to most scientists, no matter how wistfully they long for it. It is expensive folly to establish a project or institute and fill it with scientists from a variety of disciplines, selected only on the basis of scientific credentials. Without leadership, specific common goals, mutual empathy and human consideration, and a great deal of effort, the result will be a collection of scientists none of whom has a colleague. Finally, it should be noted that it is not easy to get funds for a large, interdisciplinary project. It is important to find a sponsoring agency that is willing to invest in long-term research, because continuity is critical. A team cannot be brought together for productive research and disbanded more than once. We are grateful to ARPA and NIH for providing such funding support.

CHAPTER
NINE

SUMMARY AND CONCLUSIONS

A key to DENDRAL's success is its knowledge of chemistry, mass spectrometry, graph manipulation, and other technical material. Engineering the acquisition and use of that knowledge was an enormous task. Both Dendral and Meta-DENDRAL are built on a three-stage model of hypothesis formation—plan, generate, and test—that constitutes a model of scientific discovery.

9.1 INTRODUCTION

The DENDRAL Project is a study of scientific reasoning. One major thrust of this work has been the exploration of methods for acquisition, representation, and use of knowledge. We have referred to the design of such methods as *knowledge engineering*. In this chapter we will elaborate on this theme, which is the basic *engineering* aspect of the work. An implicit second preoccupation of this work has been the collection of observations about *scientific discovery* that might be pertinent to a more systematic theory of discovery. In this chapter we will elaborate also on this theme, which is the basic *scientific* aspect of the work.

9.2 KNOWLEDGE ENGINEERING

At the time of inception of the DENDRAL Project, the major emphasis of most AI research was a search for general methods of problem solving. Relatively little effort was devoted to the design of systems that embodied and used specialized knowledge. The paradigm case of the search for general methods was research on the resolution method of proving theorems in the predicate calculus, as developed by Robinson (1965) and applied to a variety of problems [e.g., robotics; Raphael (1972)] by a number of computer scientists. There were other important elaborations of this theme, for example, the General Problem Solver [Ernst and Newell (1969)], which we discussed in Chapter 3. There were also notable exceptions, projects in which emphasis was placed on specialized knowledge (chess and checker programs are good examples, as are symbolic mathematics aids such as MACSYMA [Martin and Fateman (1971)]; such work, however, was in the minority.

The situation today is quite the opposite. The search for general methods of problem solving is no longer the mainstream of AI research, while many significant projects can best be characterized as the development and application of knowledge-based systems. Thus approximately 30 percent of the papers at the 1977 International Joint Conference on Artificial Intelligence reported applications and specialized systems. It is interesting that game-playing programs, particularly for chess, continue to thrive.

From the beginning, the DENDRAL Project pursued a strategy of encoding large amounts of task-specific information into heuristic programs, a strategy now known as "knowledge engineering." Indeed, the success and example of the Project in all likelihood played an important role in the establishment of the new emphasis in AI, though it was not the only force acting in this direction. We now are more convinced than ever that the design of knowledge-based systems is an important emphasis, and a more productive path at the moment than the search for general methods of problem solving. We do not wish to argue that general methods of problem solving are a logical impossibility. However, even if such methods are possible and attainable, they will not replace, but merely augment, systems for acquiring and using specific knowledge. With the benefit of hindsight and a mature design of the DENDRAL system at hand, it is possible to make this case more concretely.

We take it to be self-evident that problem solving in a specific task domain requires special knowledge of that task domain. This was not contested, merely not emphasized, a decade ago. In a predicate calculus–based system, specialized knowledge was encoded in the axioms, the theorem-proving procedures, and the criteria of interest; it was not ignored. In GPS, specialized knowledge was encoded into the definitions of the problem space and transformations; it was not ignored. What was not fully appreciated was the sheer *amount* and *variety* of such knowledge underlying intelligent behavior. General methods went awry when the unavoidable profusion of specialized knowledge swamped the heuristic methods and, further, outran the abilities of the programmers to encode it all. This breakdown happened as soon as attention was directed away from highly abstracted, simplified "toy" problems toward applications of utility outside AI itself.

A surprisingly large amount of specialized knowledge is needed to achieve expertise in even a very circumscribed field. The fact that long periods of time are required to become an "expert" is evidence that expertise is knowledge-intensive. For example, Simon and Barenfeld (1969) present evidence that the difference between expert and novice chess players lies almost exclusively in their differing degrees of familiarity with commonly occurring patterns of chess pieces, a familiarity reflected in speed of recognition and ability to recall, and acquired by extended experience. It is estimated that the chess expert is familiar with between 10,000 and 100,000 such patterns, which is also the range of the word-recognition vocabulary of a fluent speaker. In addition to pattern familiarity is a host of other knowledge that novice and expert alike must share, and that passes unnoticed in a casual analysis of game playing. This knowledge includes, for example, information about the nature and purpose of games and of competition generally. In the case of mass spectrometry, the knowledge base includes not only specialized knowledge of technique, but a large amount of information about the underlying subjects of chemistry and graph theory, any portion of which may profitably be brought to bear in the solution of a particular structure elucidation problem. A prerequisite of successful performance by DENDRAL was the encoding of significant portions of this knowledge base, a job that has taken thousands of man-hours.

To gain sufficient problem-solving power in the face of the needed quantity of knowledge, a knowledge representation scheme must be sufficiently specialized; making it so is a major part of the engineering problem. Consequently, successful knowledge engineering initially requires decisions about knowledge representation; in particular, specialized representations for specialized applications. A representation that is uniform for *all tasks* is doomed to impotence in problem-solving power, although we have argued previously that uniformity of representation has significant advantages, within a given task and for a given purpose. *Thus the basic representation used by DENDRAL, chemical graphs, is, we have noted, the glue that holds the system together and permits various qualities of knowledge to combine effectively.* However, chemical graphs are manifestly not the appropriate knowledge representation for chess, or for speech understanding, or even for quantum mechanics. Furthermore, even though it is conceivable though unlikely that *some* form of graph structure will suffice for encoding all knowledge (just as it is conceivable though unlikely that some linear logical calculus will suffice), it appears that the requirements of any given problem domain are so specialized that the appropriate form of graph will be in turn so specialized as to diminish seriously the importance of whatever insight such a commonality of language might hold. Therefore, knowledge representation takes on a status equal to that of heuristic exploration in the struggle against combinatorial complexity.

To elaborate: the goal of knowledge engineering is to achieve a productive interaction of knowledge in the service of problem solving. An appropriate knowledge representation is an encoding that productively relates information that is naturally related in important ways in the referent application area. Further, it ought to do so in ways that ease the burden of inference [see Lindsay (1961) and Lindsay (1973)]. To the extent that a measure of the inferential burden can be borne by the representation scheme, we have reduced the burden that must be borne by search and generation heuristics.

To summarize: successful problem solving in nontrivial domains (1) requires surprisingly large amounts of specialized as well as general knowledge, (2) requires different forms of organization for different tasks, and for different purposes within a given problem domain, (3) requires the productive interaction of this knowledge, not merely its accumulation, and (4) can benefit from representation schemes that bear part of the burden of the inferential process. For these reasons, which have the status of empirical propositions about cognitive systems generally and human minds specifically, we conclude that the current emphasis on knowledge engineering within AI, for which

DENDRAL is a key example and important case study, is both central and prerequisite to the development of artifacts of general intelligence.

9.2.1 How Much Does DENDRAL Know?

It is frequently asked *how much* DENDRAL "knows." Unfortunately there is no straightforward answer to this question. We can say, qualitatively, that DENDRAL employs a lot of knowledge, perhaps the equivalent of much of high school-level organic chemistry, plus some specialized facts from college-level and even graduate-level chemistry. However, at present no theory offers a means to quantify knowledge, even when knowledge is embodied in a form (such as a computer program) whose structure is in principle completely explicit. Many difficulties impede the development of a quantitative theory of knowledge.[1]

For example, it has frequently been pointed out that humans, and, with qualifications, computers, know in essence an infinite number of facts (to cite one instance, we know the successor of any integer); and yet our memories are finite. Therefore, quantitative theory must be able to handle inferential, generative knowledge of this sort.

Two additional major difficulties are, first, that any piece of knowledge is meaningful only in a larger context of other knowledge, and, second, that cognitive capacities can be represented in indefinitely many ways.

For example, we may assert that DENDRAL knows that the valence of carbon is 4. Knowing this fact, however, presupposes that the concepts of valence, atom, and a constellation of related concepts are also in some sense understood. (Even so, DENDRAL's concept of valence is clearly not the same as a chemist's, which is imbedded in an even richer context of related knowledge.) It is, furthermore, knowledge that is distributed throughout many subprograms that define and manipulate chemical graphs, or in some way make use of facts such as "the valence of carbon is 4." It would be nearly impossible to separate those pieces of DENDRAL computer code that in one way or another are associated with an understanding of valence from those that are not.[2]

In analyzing the issue of multiple representations of knowledge, the conventional distinction between "knowing how" and "knowing that," though itself not precise, is a helpful starting point. In the case of the majority of programmed *algorithms* (Section 3.2.1), it is often possible to distinguish processes from propositions (facts, data, and parameters), corresponding to the distinction between knowing how and knowing that. Even when this division is possible, however, it must be remembered that there exist many different but functionally equivalent, and hence equally "knowledgeable," programs that divide their knowledge in different ways between processes and facts. For example, a stored table of logarithms and a program for computing just those loga-

---

[1] While we are not prepared to propose such a theory, we can make an important terminological suggestion. In analogy to the accepted unit of information, the *bit*, we propose to call the unit of knowledge the *knit*. The unit of wisdom would then, of course, be the *purl*.

[2] Some of the programs have been more meticulously written in this regard than others. For example, INTSUM always references the valence of chemical atoms through a single function. However, the more general concept of connectivity of graphs, which subsumes valence, is part of the whole framework assumed by almost all functions.

rithms know the same amount, although the former program is based on knowing that, while the latter is based on knowing how. Further complications arise from program structures such as productions. A set of productions is a peculiar combination of knowing how and knowing that; it is a propositional description of a process, a sort of "knowing that this is how."

A count of how many facts DENDRAL knows, even if this were possible, would not be a fair description of its total knowledge, since much of its knowledge is embodied in executable code. In combining different kinds of knowledge we almost need to abandon static metrics altogether in favor of comparing the dynamic uses of the facts and procedures. For example, we might compare the times each of two bodies of knowledge takes to reach a result.

Heuristics further muddy the waters. To mention just one major problem, much heuristic knowledge is uncertain ("this method frequently works but is not guaranteed"). While information theory tells us how to quantify the *information* content of an assertion based on its a priori probability of truth, we have no comparable means of deciding the quantity of *knowledge* conveyed by probabilistic statements such as heuristics. Gaschnig (1977) explores related problems of measuring the power of heuristics. All such measurements, however, are dependent on the actual implementation of the knowledge in the program.

Finally, even knowledge that appears to be neatly parceled into individual facts cannot simply be counted. To do so meaningfully would require, first, the development of a precise calculus permitting decomposition of each complex fact into canonical form so that its components could be counted. One candidate for such a calculus would be computer machine code: amount of knowledge would be the number of compiled instructions needed to store it. For many reasons, among them the sensitivity of this measure to the seemingly irrelevant factor of compiler efficiency, this enumeration is not a satisfactory answer. Instead, the thrust of the initial question presupposes a semantically interpreted calculus of a form that corresponds to human cognitive organization. Needless to say, such a calculus is not at hand. We do not know what a maximally compact form for knowledge would be, nor would we know how to prove that it is such. We use abstractions for complex facts and procedures (e.g., macros, in programming terms) as a means for compacting the symbols used for expressing the knowledge. But we must not forget that these compact forms embody much semantic information. One of the most important issues of science is the organization of knowledge including the use of abstractions for condensing it.

For these reasons any list of what DENDRAL knows in terms of concepts, facts, and processes is of limited descriptive power. Nonetheless, we will attempt to classify the content of DENDRAL's knowledge of chemistry in the table below. [See a similar discussion of what Winograd's SHRDLU knows in Boden (1977), pages 134-142.]

DENDRAL's Knowledge of Chemical Concepts and Procedures

1. Knowledge of chemical graphs
   a. Atom types (C, H, N, O plus provision for adding others)
   b. Valence of each atom type
   c. Atomic weight of each atom type

d. Bond types (single, double, triple, aromatic)
   e. How to detect topological symmetry
   f. How to compute degree of unsaturation from empirical formula
   g. How to draw reasonable planar projections of molecular structures
   h. How to generate all isomers including fused rings, spiro forms, etc.
   i. How to generate all stereoisomers
   j. How to find cycles and arbitrarily complex subgraphs
   k. How to find the greatest common subgraph among a set of graphs
   l. How to label nodes and edges of graphs in all distinct ways, taking account of symmetry
   m. How to simulate specific chemical transformations, such as synthetic reactions
2. Knowledge of chemical stability
   a. Twenty classes of unstable acyclic structures known; any others can be specified
   b. How to recognize keto-enol tautomerism; other tautomers can be specified
   c. Terpene rule
   d. Isoprene rule
   e. Bredt's rule
3. Knowledge of mass spectrometry
   a. How to infer the formula of any molecular ion
   b. How to compute results of any specified fragmentation and rearrangement
   c. How to predict metastable peaks and use them for confirmation of inferences
   d. Rule of charge placement on fragments (but not on atoms)
   e. Half-order theory produces rough prediction of actual spectra
   f. Refined theory can be added for any family of structures (now available for ketones, ethers, alcohols, amines, thiols, thioethers, estrogenic steroids, keto-androstanes, marine sterols, and aromatic acids)
   g. McLafferty rearrangement, water elimination, carbon monoxide elimination, carbon dioxide elimination, and elimination of other user-defined "neutral species"
   h. Distinguishes high- and low-resolution spectra
   i. Distinguishes low voltage and high voltage measurements
4. Knowledge *not* available to DENDRAL
   a. Three-dimensional structure (except of stereoisomerism)
   b. Polymeric structures
   c. Quantum mechanical explanations of mass spectrometry processes
   d. Electronegativity
   e. Physical properties such as dipole moment, molecular susceptibility, melting point, crystal structure, and many others

In addition, all the knowledge of LISP is presupposed by the DENDRAL programs. For example, arithmetic and set theoretic operations, symbol manipulation, interpretation of complex procedures, and countless bookkeeping operations. Considerable amounts of code are devoted to keeping track of intermediate results in the overall processing. This "specialized bookkeeping" knowledge is not very profound, yet it is indispensable for the integration of many complex procedures.

Almost all DENDRAL's knowledge is tailored to the task of molecular structure elucidation. In spite of the elegance and simplicity of computing concepts we have to work with, the problem-solving procedures in DENDRAL are still very special purpose, complex, and voluminous. Making the procedures, and knowledge base, more general would have increased the burden of debugging them in most cases.[3]

9.2.2 How Is Knowledge Employed in Heuristic DENDRAL and Meta-DENDRAL?

In Chapter 3 we described the organization of problem-solving systems. The major bifurcation was between algorithms and heuristic programs. Heuristic programs were further characterized as either search through a *space of subproblems* or as generation of candidates from a *space of potential solutions*. It is also possible that a problem space consists of partial solutions, as in the HEARSAY speech-understanding system, [Erman and Lesser (1978), among others]. A space of partial solutions is in effect a combination of a space of solutions and a space of subproblems.

In either case, search or generation, the alternatives considered may be limited to those known to be legal according to a given rule, or to those that are merely plausible. In each case, problem solving may terminate with the discovery of a satisfactory solution, or proceed until the optimal solution is found.

To this initial division of heuristic programs, we added the concept of planning as exemplified by the planning phase of DENDRAL's basic solution generation method. More generally, the use of planning can lead to computational economies in two ways. Planning can *prune* the space (of subproblems or solutions) by eliminating sections of it and directing search to certain other sections, or it can *guide* the problem solving, either by ordering the search/generation sequence, or by modifying one subproblem/solution candidate to produce the next, using a hill-climbing (evolutionary) method. In the hybrid case of a space of partial solutions, both pruning and guidance methods are applicable.

Planning, furthermore, may be characterized along a different dimension. It may be data-driven or expectation-driven. Data-driven planning begins by examination of data (from instruments, perhaps) and attempts to induce hypotheses to account for it. Expectation-driven planning begins with a model of the phenomenon and uses it to establish expectations.

It is possible to establish a taxonomy of problem-solving systems based on these four binary characteristics: (1) subproblem space versus solution space, (2) legal versus plausible alternatives considered, (3) pruning heuristics versus guidance heuristics, and (4) data-driven versus expectation-driven. While such a taxonomy is useful, it should be remembered that the values of these dimensions are not mutually exclusive nor exhaustive, and indeed greater power will probably derive from future systems that combine these methods in various ways.

Heuristic DENDRAL may be characterized as a *generator* of *legal* solutions with

---

[3] For these reasons, we look forward to advances in automatic programming that will simplify the programming, debugging, and interfacing of complex procedures.

pruning by *expectation-driven* heuristics. Meta-DENDRAL may be characterized as a *generator* of *plausible solutions* with *pruning* by both *expectation-driven* and *data-driven* heuristics. Within this design remain several variations. For example, the DENDRAL generator is based upon a canonical form of hypothesis, and is nonredundant and exhaustive. These are desirable features but will not be available for all applications. However, if a particular application does not yield a generator with these properties, planning may still be possible, and the system can achieve considerable power simply from a systematic application of knowledge.

9.3 SCIENTIFIC DISCOVERY

Meta-DENDRAL evolved in a natural way from Heuristic DENDRAL. The concept of a generator of hypotheses is particularly transparent in the case of Heuristic DENDRAL, and the hypotheses themselves were simple in structure. The understanding achieved from the Heuristic DENDRAL system made possible the evolution of Meta-DENDRAL, in which the hypotheses took on a complexity much closer to that of the usual conception of a scientific hypothesis. We now feel that we have gained some insight into the larger issues of scientific discovery in the Baconian spirit of rational directions to the uninitiated for formulating scientific hypotheses.

9.3.1 Historical Background

There is a substantial literature on the problem of induction (briefly, how one may pass from particular statements to universal statements), which, following Bacon, has traditionally been proposed as the basis of scientific discovery, and is so presumed to be in ordinary discourse. However, most contemporary philosophers of science disagree with this view. For example, Medawar (1969) discusses induction at length and concludes that it is not the basis of scientific discovery. He argues, as have other recent writers, that science proceeds by a hypothetico-deductive scheme, that is by discovery—the invention of theories—followed by verification—the comparison of theoretical predictions with empirical observations. These two components pose two basic questions for the philosophy of science. The first is the *problem of discovery*, that is, how hypotheses arise. The second is the *problem of verification*, that is, the relation of a given hypothesis to evidence.

The question of scientific discovery, when not ignored altogether, is frequently relegated to psychology, which has yet to embrace it. There is no disciplined effort to collect well-documented narratives of discovery in science. The result is that almost nothing has been established about the nature of scientific discovery other than a collection of poorly documented anecdotes about how ideas "arise" in scientists' mind, often while they are engaged in nonscientific pursuits: the eureka hypothesis. For example, Medawar characterizes the discovery process as "an imaginative preconception of what might be true" (p. 51) that is mediated by "inductive intuition": ". . . thinking up or hitting on a hypothesis from which whatever we may wish to explain will follow logically" (p. 56). This process he calls the generative act of scientific discovery and is bold enough to suggest "That 'creativity' is beyond analysis is a romantic illusion we must now outgrow" (p. 57).

Some writers have indeed harbored this "romantic illusion," dismissing the possibility of any serious study of the question of scientific discovery. For example, Popper (1968, p. 31) states, in a book entitled *The Logic of Scientific Discovery*:

> I said above that the work of the scientist consists in putting forward and testing theories.
> The initial stage, the act of conceiving or inventing a theory, seems to me neither to call for logical analysis nor to be susceptible of it. The question how it happens that a new idea occurs to a man—whether it is a musical theme, a dramatic conflict, or a scientific theory— may be of great interest to empirical psychology; but it is irrelevant to the logical analysis of scientific knowledge.

9.3.2 The Plan-Generate-Test Model of Discovery

Our central point here is to elaborate the idea that scientific discovery can be profitably viewed as a systematic exclusion of hypotheses. This view is another instantiation of the plan-generate-test paradigm. The conditions under which this view makes sense are an important part of the elaboration. Two necessary conditions are that the space of relevant hypotheses is definable, and that criteria of rejection and acceptability exist. Because the space of hypotheses is immense for most interesting problems, it is also desirable that criteria exist for guiding the systematic search.

The method of proof by eliminative induction, advanced by Bacon and Hooke, was dropped after Condillac, Newton, and LeSage argued successfully that it is impossible to exhaustively enumerate all the hypotheses that could conceivably explain a set of events [Laudan (1973)]. The method advanced in our work is in some sense a revival of those old ideas on induction by elimination, but with machine methods of generation and search substituted for exhaustive enumeration. Instead of enumerating all sentences in the language of science and trying each one in turn, a computer program can use heuristics enabling it to discard large classes of hypotheses and search only a small number of remaining possibilities. However, a high price may be paid for reducing the search: sometimes the inquirer will not see the best solution, or perhaps may not see any solution.

9.3.2.1 The source of scientific hypotheses

Kuhn (1962), in his well-known analysis of the history of science, introduces the concept of a scientific paradigm. Before the establishment of a paradigm, a science is in a presystematic, natural history stage, characterized by competing theories and emphases. With the ascendence of a theory that gains wide acceptance—the paradigm—research is of a different type, which Kuhn calls normal science. This stage is characterized by general agreement as to what are the key conceptual problems and accepted experimental techniques, and what empirical data demand explanation. Discovering hypotheses in the natural history stage of a science is not like discovering hypotheses in the more advanced theoretical stages of the science. (But it should be remembered that these stages may be found in one science at the same time, and that they are not clearly separable.) In the natural history stage the discovered hypotheses are often universal generalizations about observables. These hypotheses serve less as explanations than as descriptions of regularities in the universe; one might well ask why the generalization holds, thus asking for a higher-level statement to explain the observed regularity, that is, for a theory in terms of which the regularity can be explained. On the level of descriptive science, the inductive generalization is a reasonable discovery to expect, but on the level of explanatory theory we expect more than a description of what is the case; we expect to discover hypotheses that explain why the lower-level generalizations hold.

As a result of distinguishing descriptive science from explanatory science, it would seem that simple inductive inferences, using the vocabulary of a fixed model, may lead to discoveries of useful hypotheses in the descriptive stage but not in the more advanced, theoretical stage. Also it becomes apparent that in the two different stages there are different criteria for what will count as a useful hypothesis. The purpose of hypotheses in the natural history stage of science are mainly for description and classification, but in the explanatory stage they are to explain phenomena and unify diverse explanations. This separation of the descriptive from the explanatory aspects of scientific inquiry also suggests that there are probably different logics of discovery for these activities.

In general, the problem for a logic of discovery is twofold: to choose a language L in which to express hypotheses explaining data in a scientific domain, and to choose a satisfactory sentence of L that explains the data. In paradigm revision, the first half of the problem is crucial, for the choice of the language establishes boundaries on the factual content of the paradigm. Choosing to speak of light as traveling, to use an example suggested by Toulmin (1961), determines in large measure the kinds of questions we ask about light and the kinds of answers the paradigm will furnish. On the other hand, in normal science choosing a language is not part of the problem, for the language in which hypotheses are expressed is just the language of the current paradigm. That is, once a paradigm is established, normal scientists describe their work within that language; only when description and explanation within the language of normal science fail do scientists again face the problem of choosing a new language in which to express their hypothesis.

Although we have not learned how to engineer the formulation of new vocabularies, we believe that combinatorial play at various levels of abstraction is the key to thinking about it. We disagree with the widely held view that there is no method underlying the creation of new terms in the language of science.[4]

The second half of the problem—choosing one of the sentences of L to serve as a hypothesis—is a problem that the DENDRAL program addresses. The problem, in general terms, is to find efficient methods for picking out sentences of L that are most likely to succeed as hypotheses in a given class so that the inefficient process of enumerating sentences and trying one can be avoided. It is clear that scientists do not resort to an enumeration and one-by-one trial of sentences of L, for we would expect little or no progress in science with such inefficiency.

[4]For example, Bronowski (1966, p. 6) clearly states this view: "We do not know; and there is no logical way in which we can know . . . The step by which a new axiom is added cannot itself be mechanized. It is a free play of the mind, an invention outside the logical process."

SUMMARY AND CONCLUSIONS 163

We could consider discovery to be merely successful guessing, as is often suggested, and program a machine to perform random generation of hypotheses—perhaps restrained within the correct subject area. It could test each random hypothesis against the criteria of success and stop when a hypothesis met the criteria of reasonableness. Although some inquiring minds may work in this random manner, it hardly recommends itself as a rational method.

The trial-and-error method might not be as irrational as first appears, however, for the trials can be selected by other than random means. As Polya (1954, p. 26) sees this method:

> ... it consists of a series of trials, each of which attempts to correct the error committed by the preceding and, on the whole, the errors diminish as we proceed and the successive trials come closer and closer to the desired final result.

Because the method does not rely on random trials, Polya prefers calling this the "method of successive approximation." He notes the widespread use of the method in mathematics and from such tasks as finding a word in a dictionary to proposing scientific theories that are better and better explanations of phenomena. Using this progressive warming (hill-climbing) method, however, depends on (1) seeing what is wrong with any guess and (2) seeing how to correct it. If either of these conditions is not met, then the method does degenerate into random guessing, for the inquirer (either human or machine) has no direction for further guessing. Moreover, the techniques for performing (1) and (2) must be highly sophisticated in order to transform an initial guess into a solution of the problem in a reasonable amount of time. For instance, hundreds of things may be wrong with one of the initial guesses at a solution, but the techniques for (1) should point to the error whose correction will most advance our progress toward a solution. And, correspondingly, the techniques for (2) should correct the error in the way that most closely approximates a solution.

Hill-climbing methods to reach scientific truths from approximation were perhaps first advocated by Hartley and LeSage in the eighteenth century [Laudan (1973)]. Borrowing from the success of the mathematical method of successive approximation, they argued that positing hypotheses and correcting them will lead to truths in science. Unfortunately, they did not specify the technique for modifying or replacing false hypotheses. Even through the nineteenth century, as Laudan (1973, p. 285) notes, "Everyone assumes that science is self-corrective (and thereby progressive), but no one bothers to show that any of the methods actually being proposed by methodologists are, in fact, self-corrective methods."

Another method we could consider using is means-ends analysis, as used successfully in several computer programs, most notably the General Problem Solver (GPS). As we saw earlier (Section 3.2.2.1), this method is also a form of hill climbing. The procedure may be applicable for transforming the data into an explanation and for systematization of the data. Reducing the differences would be a powerful tool for hypothesis formation if the conditions that define "explanation and systematization" can themselves be made precise.

9.3.2.2 Guiding the consideration of hypotheses Philosophers of science have not framed the problem of hypothesis formation and testing to include formulation of constraints on acceptable hypotheses. We suggest that it is important to do so. That is to say, the planning stage of the process is of key importance.

9.3.2.3 Checking the hypothesis Another necessary condition for any successful problem-solving activity is that a solution to the problem can be recognized if it is found; that is, that there be well-defined boundaries on what will count as a solution to the problem.

In the search for a reasonable hypothesis for some purpose, the inquirer must, then, have some criteria by which to judge when the search terminates successfully. And for this reason, criteria of reasonableness must be made precise before a logic of suggestion can be developed. Thus, all the problems of making the criteria precise carry over to the search for methods of hypothesis formation; without the criteria the methods could not be developed systematically.

9.3.3 Outline of a Model of Scientific Discovery

Embodied in both the Heuristic DENDRAL and the Meta-DENDRAL systems *is* a method of discovery. We here propose that this method might underlie scientific discovery in other areas and suggest that its basic characteristics have the status of a theory of scientific discovery, though possibly incorrect and clearly incomplete (as measured against the suggestions we have just discussed). The basic claims of such a theory are the following.

1. Scientific discovery uses the same basic methods of problem solving as do other scientific reasoning and other forms of problem solving.

We note that Meta-DENDRAL is not much different in organization from Heuristic DENDRAL. Both are a species of plan-generate-test. Both are guided by a strong model of the domain, although Meta-DENDRAL does coarse search first. It was a surprisingly small step from Heuristic DENDRAL to Meta-DENDRAL. No basic additions or reorganizations of the problem-solving method were required, even though in the latter case the results of problem solving are hypotheses (in the form of productions) that embody a limited scientific theory (concerning the behavior of a class of compound in the mass spectrometer).

2. Scientific discovery is judicious selection from a space of possible hypotheses by heuristic exploration.

A cognitive agent has means for generating either the members of a set of possible hypotheses or the states of a problem space. This ability is productive in the sense that human language is productive (as discussed by linguist Noam Chomsky): a large set of novel combinations of a finite set of elements can be generated even though none has been previously encountered in the experience of the scientists.

3. In exploring the space of possible hypotheses, the scientist is strongly influenced by initial assumptions.

Heuristic DENDRAL is a theory-driven mechanism: it finds only what it is looking for. For example, if it does not expect to find solutions (organic compounds) containing certain substructures, as signified by the presence of these substructures on BADLIST, then they are not generated. It is just such biases, (and the more the better), that allow the programs to discover a manageable set of candidates at all.

The question arises from whence these biases issue when they take the form of predisposition toward certain types of *theories*. The answer in the case of Meta-DENDRAL is that they arise from presystematic notions about mechanisms underlying the phenomena of MS behavior, as derived from yet other theories: the scientific paradigm of the discipline. The particular forms that the generated theories take is initially dictated by observations (data) and refined by interactions of data and criteria of explanatory power. We have nothing to say about the origins of new paradigms; the nature of the ways in which the initial forms arise and are refined are the subjects of items 4 and 5.

4. Scientific problem solving in general, and discovery in particular, outside the well-codified areas of science, involves the employment of a large number of vague and unverified ideas, rather than the application of logical deduction to previously verified propositions.

It is clear that Heuristic DENDRAL does not have at its core a formal theory of chemical stability, but rather employs a large collection of weak partial assumptions each of limited range. Nor does Meta-DENDRAL have at its core a formal metatheory of MS theories. We propose that this lack of formal theory is the rule in scientific discovery, in contrast to the classical description of scientific method as tight reasoning from established premises. Again, contemporary writers have expressed a similar opinion; what DENDRAL contributes is substantive detail in elaboration of this view.

5. Scientific problem solving in general, and discovery in particular, is an interaction of top-down (expectation-driven) exploration, and bottom-up (data-driven) exploration. Both are necessary.

Heuristic DENDRAL is driven largely in a top-down manner. However, the data (spectra) are employed by MOLION in planning and after the hypotheses are at hand, when PREDICTOR attempts to winnow the set of candidate solutions. Meta-DENDRAL by contrast is driven in both directions. First the spectra are used, by INTSUM, to establish the hypothesis space. The hypothesis space is then used, by RULEGEN, to produce candidate solutions (hypotheses). Finally, the data are again used to refine the set of hypotheses into a more manageable, parsimonious theory.

6. A generator that fails to guarantee completeness is not wholly satisfactory, since one then cannot say with certainty what hypotheses have been excluded from consideration.

7. A generator that cannot avoid duplicate (or equivalent) expressions of the same hypothesis is not wholly satisfactory, since the generation may never terminate.

8. Knowledge employed early in the exploration constrains the search more efficiently than knowledge employed later.

9. Knowledge about *classes* of hypotheses is more effective than knowledge about *individual* hypotheses.

10. A small set of plausible alternative hypotheses resulting from the generation and testing may be as valuable as a single hypothesis.

The upper bound on the acceptable size of the found set varies with problem complexity and with the ease of discriminating among the alternatives by other means.

Aside from the plausibility of this theory that inheres in the existence of successful programs, no specific data support it. The theory itself suggests where we might look for support, of course, since we now have it as a source of expectations. There is anecdotal support in abundance. If the normal mode of processing empirical data is to verify expectations, as our theory has it, then science should appear essentially conservative, that is, exhibit few novel theoretical formulations. Kuhn (1962) argues this point at length. However, it is not our intent here to argue the validity of this theory, only its plausibility. The proposed elements characterize a theory of scientific discovery that, we have noted, has few competitors. The details of the DENDRAL programs as presented in the earlier chapters richly illustrate ways of making this theory specific and suggest numerous alternative specifications.

As an example of one particular form such a theory might assume, we present the following description that follows in some detail the existing architecture of Meta-DENDRAL.

1. Presupposed is a paradigm that provides a set of concepts deemed relevant and important, a model of their interrelationships, and a generative grammar that defines the ways in which these elements may be meaningfully combined to state hypotheses.
2. Empirical observations, assumed to be sufficiently reproducible, are selectively examined by looking at those features of them that the paradigm suggests are significant. These abstracted descriptions of the data are examined by simple pattern-recognition processes to detect important regularities.
3. Separate hypotheses are constructed to explain each detected regularity by generation (using the grammar described by the paradigm) guided by planning (based on limitations of complexity also proposed by the paradigm).
4. The collected set of hypotheses is examined and reduced by the elimination of those that make small or redundant contributions to the account of the data.
5. The set of hypotheses is modified by generalizing each in turn, and specializing each in turn, in an iterative procedure that refines the set under control of the nature and quantity of correct and incorrect predictions made by the changing set of hypotheses, until criteria of generality and simplicity are met.
6. At each of the above steps, highly task-specific knowledge is employed in varied ways as heuristic methods of search, as stopping rules, and as measures of goodness of fit. Any particular theory of scientific discovery must perforce be domain-specific in large measure.

Under our analysis the traditional problem of finding an effective method for discovering true hypotheses that best explain phenomena has been transformed into finding heuristic methods that generate plausible explanations. The problem of giving rules for producing true scientific statements has been replaced by the problem of finding efficient heuristic rules for culling the reasonable candidates for an explanation from an appropriate set of possible candidates.

In the most creative heights of science, hypothesis formation is farthest from the "reach of method" as Whewell (1858) says. But within the comfort of an established scientific theory, paradigm, or conceptual scheme, hypothesis formation usually does not involve the introduction of new concepts. The concepts are given and the task of a logic of suggestion is to show how hypotheses should be formulated in terms of these concepts. Depending on the purposes at hand, and in part on the science, the hypothesis may either explain a puzzling phenomenon (or set of phenomena) or describe objects and events within the scope of the science.

The problems with formulating this kind of logic of discovery are both difficult and numerous. Before any methods, heuristic or otherwise, can be given for "discovering" explanations or regularities, a precise conception of success for the logic must be formulated. When criteria are clarified and refined for the specific science considered, then the methods could be said to succeed when they produce hypotheses that meet the criteria. The methods themselves will also be difficult to formulate in specific instances because of the difficulties in understanding the problem, representing the space of possible solutions, dividing the task into subproblems, and planning a solution, to mention the outstanding ones.

To a modest degree, the Heuristic DENDRAL and Meta-DENDRAL programs capture many of the notions of a logic of discovery. They are more systematic and less random than we have come to expect of creative guessers in science, but their methods are also more teachable and their results more reproducible.

9.3.4 Limitations of Computer-Aided Discovery by Heuristic Search

The major limitation of the heuristic search method in any domain is the necessity of finding (or inventing) a generator of possible solutions. In the rule-formation domain, that necessity means that we have to invent a program that generates possible rules. That, in turn, requires a strict definition of the allowable forms of the rules and a definition of the allowable primitive terms that add content to the form. The representation we have found for expressing rules is fixed for any one run, but can, at least, be modified or extended manually between runs. In the case of molecular structures, finding the generating algorithm took many years. Lederberg's notational algorithm for unringed graph structures was mapped into a generating algorithm with little difficulty, but the symmetries of cyclic graphs complicated the generation problem immensely. Not until considerable mathematical expertise had been focused on the problem was a generator invented that carried guarantees of complete and nonredundant generation.

A second major limitation on heuristic search is the necessity of finding heuristics, rules of thumb, that guide the generator and constrain it from producing all syntactically allowable hypotheses. For rule generation it is necessary to find heuristics that steer the generator toward the small number of interesting rules and away from the very large number of uninteresting rules. The problem is that it is difficult to find these guiding principles. In addition, putting confidence in the heuristics requires an act of faith. Once that step is made, however, the temptation is often to put *too much* faith in the heuristics and forget that the solutions were found in the context of a large number of assumptions. For example, one might tend to forget the criteria for data filtering, or the range of allowable hydrogen transfers, or the restrictions on how complex the rules were allowed to become, or the criteria for filtering the rules. All the heuristics together define the range of rules considered and thus should temper our judgments about the generality of the rules.

Another limitation on the use of heuristic search is that the computer programs are often slow, not because they are inefficient so much as that they must do a lot of computation. The Meta-DENDRAL program is also inefficient now because it is still in the development stage. This practical difficulty limits the sizes of the problems the programs can solve now. Since CONGEN can readily solve some problems complex enough to challenge and interest human chemists (as can Meta-DENDRAL, with some difficulty), the method at least has proved feasible.

Limitations are also imposed by the domain of chemistry, which have been mentioned elsewhere. To reiterate, the programs work with a connectivity model of chemical structures, without knowledge of geometric properties. Nearly everything that a chemist can tell the programs needs to be expressed in terms of subgraphs. Also, the programs depend on a chemist's judgment for their chemical heuristics. (This dependence is also a strength as well as a limitation.)

9.4 THE PROSPECTS FOR AUTOMATIC SCIENCE

DENDRAL illustrates the state of the art in automatic hypothesis formation. It can lay claim to this position not merely because it has few competitors at the moment, but because it has been a thorough, sustained effort by an interdisciplinary group of scientists. The scope and power of the program are therefore a good indication of what can be done with the technology of today. Our forecast for the immediate future is optimistic for projects of modest scope, attacking well-defined scientific problems in a manner that allows the full power of the human mind to be augmented by the complementary powers of the computer.

How far this endeavor can be carried and what ways it will change in the future we can only guess. We realize that proposing significant mechanization of scientific thought, or of any cognitive ability that to date has been uniquely human, is controversial and problematic and leaves one open to accusations of committing the sin of pride. But surely, since man did not design man, it is no more prideful to suppose that the human mind was created to be sufficiently percipient, sentient, and conscious to understand itself than to suppose that it is not. Our reverence for the human mind is undiminished, indeed is enhanced, as we explore it more deeply. We recognize that the bases for its abilities and the boundaries of its potential are yet shrouded in mystery. Nonetheless, we remain impressed by the rapid growth of knowledge and technology and are tempted to extrapolate it beyond our vision. From Pascal, Babbage, and Turing to the hand-held, programmable, microsecond computer of today is a step of awesome compass. Yet the human mind has been neither replaced nor enslaved, but freed for grander enterprises. That is a prospect with which we can live.

CHAPTER
TEN

PROJECT PUBLICATIONS

This chapter contains a partially annotated, *chronological* list (by date of writing) of DENDRAL Project publications. A complete, *alphabetized* list of references for the citations in the text follows. That list includes all items from this list of Project publications, without annotations.

1. Lederberg, J. *Computation of molecular formulas for mass spectrometry.* San Francisco: Holden-Day, 1964.
   A brief text in the Holden-Day series on physical techniques in chemistry. Presents mass spectrometry as a useful tool for organic chemists, and introduces procedures for simplifying the analysis of mass spectrometry data. Intended for calculations performed without the aid of a computer.
2. Lederberg, J. DENDRAL-64, a system for computer construction, enumeration, and notation of organic molecules as tree structures and cyclic graphs, part I. Notational algorithm for tree structures. *Report No. CR-57029* and *STAR No. N65-13158.* National Aeronautics and Space Administration, 1964.
   The first of a series of three technical reports to NASA. Introduces the DENDRAL notation and the DENDRAL algorithm for generating all the structural isomers of a given formula. Deals with those chemical graphs that are pure trees.
3. Lederberg, J. DENDRAL-64, a system for computer construction, enumeration and notation of organic molecules as tree structures and cyclic graphs, part II. Topology of cyclic graphs. *Report No. CR-68898* and *STAR No. N66-14074.* National Aeronautics and Space Administration, 1965.
   The second of three technical reports to NASA. Introduces the notion of Hamilton circuits as a scheme for constructing canonical names of cyclic graphs. This notational algorithm did not bear fruit as a generating algorithm.
4. Lederberg, J. Topological mapping of organic molecules. *Proceedings of the National Academy of Sciences,* 1965, 53:1, 134-139.
   Summarizes the notions introduced in the series of technical reports to NASA [Lederberg (1964b), (1965b), and (1970)]. A clear and nontechnical presentation of an algorithmic approach to the topological mapping of organic molecules.

5. Lederberg, J. Systematics of organic molecules, graph topology and Hamilton circuits, a general outline of the DENDRAL system. *Report No. CR-68899* and *STAR No. N66-14075*. National Aeronautics and Space Administration, 1966.

A general introduction to the DENDRAL system for chemical structure notation; an introductory survey of the notions underlying that notation.

6. Lederberg, J. Hamilton circuits of convex trivalent polyhedra (up to 18 vertices). *American Mathematical Monthly*, 1967, 74:5, 522–527.

The first six papers contain the seminal ideas of the DENDRAL project. This paper discusses procedures for finding Hamilton circuits and presents an algorithm for finding Hamilton circuits of acyclic graphs.

7. Sutherland, G. L. DENDRAL—A computer program for generating and filtering chemical structures. *Stanford Artificial Intelligence Project Memo No. 49*. Stanford, Calif.: Stanford University, Computer Science Department, 1967.

The first paper to discuss the DENDRAL computer program, in an early version. A fairly technical paper intended for the computer science or the chemistry audience. Gives the specific details of the subprograms that constitute this version of DENDRAL. Contains a BADLIST but as yet no GOODLIST. This version has been superseded by later work.

8. Lederberg, J., and E. A. Feigenbaum. Mechanization of inductive inference in organic chemistry. In B. Kleinmuntz (Ed.), *Formal representation of human judgment*. New York: Wiley, 1968, 187–218. Also *Stanford Artificial Intelligence Project Memo No. 54*. Stanford, Calif.: Stanford University, Computer Science Department, 1967.

A summary description of the Heuristic DENDRAL program from the standpoint of artificial intelligence research. A general discussion of the notions involved in the DENDRAL approach; does not presuppose much technical chemistry knowledge. (The program now includes a GOODLIST.) A very clear discussion and evaluation of the DENDRAL approach.

9. Feigenbaum, E. A., J. Lederberg, and B. G. Buchanan. Heuristic DENDRAL: A program for generating explanatory hypotheses in organic chemistry. In B. K. Kinariwala and F. F. Kuo (Eds.), *Proceedings of the Hawaii International Conference on System Sciences*. Honolulu: University of Hawaii Press, 1968, 482–485.

A brief general description of Heuristic DENDRAL for the artificial intelligence community. Gives an overview of the selection of the problem area, the data for the program, and its capabilities at an early stage of development.

10. Lederberg, J. On line computation of molecular formulas from mass number. *Report No. CR-94977*. National Aeronautics and Space Administration, 1968.

A brief technical report to NASA describing a program written at Systems Development Corporation for computing molecular formulas from mass number. Primarily announcing the existence and value of this program.

11. Buchanan, B. G., G. L. Sutherland, and E. A. Feigenbaum. Heuristic DENDRAL: A program for generating explanatory hypotheses in organic chemistry. In B. Meltzer and D. Michie (Eds.), *Machine intelligence* 4. Edinburgh: Edinburgh University Press, 1969, 209–254. Also *Stanford Artificial Intelligence Project Memo AI-62*. Stanford, Calif.: Stanford University, Computer Science Department, 1968.

The first description of a complete Heuristic DENDRAL program, using the acyclic generator. Includes a description of the DENDRAL notation and the algorithm for generating acyclic structures. Readily readable paper directed to the nonchemistry audience. Includes much of the special-purpose chemistry information relating to certain classes of compounds that were used in what was called the preliminary inference maker. (NOTE: The earlier descriptions of the program were formulated before the plan-generate-test paradigm was explicitly stated, so the program described here is broken down into a different set of structures. The correspondences of the earlier versions to the plan-generate-test paradigm are as follows: Preliminary Inference Maker and Data Adjuster correspond to the PLANNER; the Structure Generator corresponds to the GENERATOR, and Predictor and Evaluation function corresponds to the PREDICTOR.)

12. Churchman, C. W., and B. G. Buchanan. On the design of inductive systems: Some philosophical problems. *British Journal for the Philosophy of Science*, 1969, **20**, 311–323.

    A discussion of the problem of designing a system for performing induction; directed to the audience of professional philosophers. Views the DENDRAL program as a specific case study in the systems approach to the design of inductive systems.

13. Lederberg, J., G. L. Sutherland, B. G. Buchanan, E. A. Feigenbaum, A. V. Robertson, A. M. Duffield, and C. Djerassi. Applications of artificial intelligence for chemical inference, I. The number of possible organic compounds. Acyclic structures containing C, H, O, and N. *Journal of the American Chemical Society*, 1969, 91:11, 2973–2976.

    Paper I in a continuing series of publications for chemists on the DENDRAL project, called "Applications of artificial intelligence for chemical inference." Describes the use of the DENDRAL generator in constructing the total number of possible acyclic structures of C, H, N and O. Illustrates the use of GOODLIST and BADLIST; gives examples of the linear notation used and a summary of the results. Presents the generator as a means of defining the scope and boundaries of organic chemistry problems.

14. Duffield, A. M., A. V. Robertson, C. Djerassi, B. G. Buchanan, G. L. Sutherland, E. A. Feigenbaum, and J. Lederberg. Applications of artificial intelligence for chemical inference, II. Interpretation of low-resolution mass spectra of ketones. *Journal of the American Chemical Society*, 1969, 91:11, 2977–2981.

    Paper II in the series of DENDRAL publications for chemists. An application of the program described in Buchanan, Sutherland, and Feigenbaum (1968) to the ketones. Describes the Heuristic DENDRAL program in the form of the Preliminary Inference Maker, Predictor, Structure Generator, and Scoring Function, applied to the interpretation of low-resolution mass spectra of ketones. Program limited to monofunctional acyclic structures. Gives a general introductory description of the DENDRAL approach and a detailed description of some of the heuristics embodied in the program.

15. Feigenbaum, E. A. Artificial intelligence: Themes in the second decade. In A. J. H. Morrell (Ed.), *Information Processing 68, Proceedings of IFIP Congress 1968*. Amsterdam: North-Holland, 1969, volume II, 1008–1023. Also *Stanford Artificial Intelligence Project Memo AI-67*. Stanford, Calif.: Stanford University, Computer Science Department, 1968.

    A survey of artificial intelligence research over the period 1963–1968. Discusses the topics of heuristic programming, problem solving, and closely related learning models, and the problem of representation for problem-solving systems. Includes a description of Heuristic DENDRAL as a representative endeavor.

16. Lederberg, J. Topology of molecules. In the National Research Council's Committee on Support of Research in the Mathematical Sciences (COSRIMS), *The mathematical sciences—A collection of essays*. Published for the National Academy of Sciences (National Research Council). Cambridge: M. I. T. Press, 1969, 37–51.

    A published form of the first six papers, summarizing the notions introduced there. A detailed but not highly technical presentation. This paper introduces the notion of vertex graphs and is an attempt to extend the acyclic DENDRAL notation scheme to cyclic structures.

17. Lederberg, J., G. L. Sutherland, B. G. Buchanan, and E. A. Feigenbaum. A heuristic program for solving a scientific inference problem: Summary of motivation and implementation, In R. Banerji and M. D. Mesarovic (Eds.), Theoretical approaches to non-numerical problem solving. New York: Springer-Verlag, 1970.

    Describes Heuristic DENDRAL as a study of scientific inference making. Discusses the performance level of the program and indicates further areas of development. A nontechnical paper.

18. Schroll, G., A. M. Duffield, C. Djerassi, B. G. Buchanan, G. L. Sutherland, E. A. Feigenbaum, and J. Lederberg. Applications of artificial intelligence for chemical inference, III. Aliphatic ethers diagnosed by their low-resolution mass spectra and nuclear magnetic resonance data. *Journal of the American Chemical Society*, 1969, 91:26, 7440–7445.

    Paper III in the series of DENDRAL papers for chemists. The application of the program to aliphatic ethers. The program incorporates the use of nuclear magnetic resonance data in the Predictor as a final filter on the structures generated.

19. Sutherland, G. L. Heuristic DENDRAL: A family of LISP programs. *Stanford Artificial Intelligence Project Memo AI-80*. Stanford, Calif.: Stanford University, Computer Science Department, 1969.

   Describes the Heuristic DENDRAL program as an application of the programming language LISP. Directed to the computer science audience. Emphasizes the nonchemistry aspects of the program, particularly the generation of all tree graphs of a collection of nodes. Concentrates on indicating how automation of the DENDRAL algorithm was acomplished, in the form of the structure generator. Includes the addition of the Preliminary Inference Maker, the Predictor, and an Evaluator called "Scoring Function."

20. Buchanan, B. G., G. L. Sutherland, and E. A. Feigenbaum. Rediscovering some problems of artificial intelligence in the context of organic chemistry. In B. Meltzer and D. Michie (Eds.), *Machine intelligence* 5. Edinburgh: Edinburgh University Press, 1970, 253–280. Also *Stanford Artificial Intelligence Project Memo AIM-99*, under the title "Toward an understanding of information processes of scientific inference in the context of organic chemistry." Stanford, Calif.: Stanford University, Computer Science Department, 1969.

   A paper for the artificial intelligence audience. Describes Heuristic DENDRAL in the format of the Preliminary Inference Maker, Structure Generator, Predictor, and Tester. Presents and analyzes a lengthy sample session of eliciting chemistry knowledge from an expert for the Predictor's theory of mass spectrometry. Discusses the design problems faced in writing DENDRAL. A generally nontechnical paper that gives a good discursive view of the working DENDRAL project.

21. Buchs, A., A. M. Duffield, G. Schroll, C. Djerassi, A. B. Delfino, B. G. Buchanan, G. L. Sutherland, E. A. Feigenbaum, and J. Lederberg. Applications of artificial intelligence for chemical inference, IV. Saturated amines diagnosed by their low resolution mass spectra and nuclear magnetic resonance spectra. *Journal of the American Chemical Society*, 1970, 92:23, 6831–6838.

   Paper IV in the series of DENDRAL papers for chemists. Application of the program to saturated amines (nitrogen compounds). Also used nuclear magnetic resonance data but in the planning stage instead of the filtering or testing stage. Only the Preliminary Inference Maker is used in this program. It was not necessary to generate structures, since the Preliminary Inference Maker (using mass spectrometry and nuclear magnetic resonance data) sufficiently constrained the set of possible compounds so as to yield only the correct ones.

22. Lederberg, J. DENDRAL, a system for computer construction, enumeration and notation of organic molecules as tree structures and cyclic graphs, part III. Complete chemical graphs; embedding rings in trees. Technical report, National Aeronautics and Space Administration, 1970.

   The third of three technical reports to NASA, dealing with complete structures. Written to facilitate the programming of DENDRAL for cyclic structures; deals with some of the problems faced by a computer generator program.

23. Sheikh, Y. M., A. Buchs, A. B. Delfino, G. Schroll, A. M. Duffield, C. Djerassi, B. G. Buchanan, G. L. Sutherland, E. A. Feigenbaum, and J. Lederberg. Applications of artificial intelligence for chemical inference. An approach to the computer generation of cyclic structures. Differentiation between all the possible isomeric ketones of composition $C_6H_{10}O$. *Organic Mass Spectrometry*, 1970, 4, 493–501. Part V in the series "Applications of artificial intelligence for chemical inference."

   Paper V in the series of DENDRAL papers for chemists. Describes the first implementation of a cyclic structure generator. The approach is no longer used because it proved to be insufficiently general.

24. Buchs, A., A. B. Delfino, A. M. Duffield, C. Djerassi, B. G. Buchanan, E. A. Feigenbaum, and J. Lederberg. Applications of "artificial intelligence" for chemical inference, VI. Approach to a general method of interpreting low resolution mass spectra with a computer. *Helvetica Chimica Acta*, 1970, 53:6, 1394–1417.

Paper VI in the series of DENDRAL papers for chemists. Important use of mass spectrometry theory to generate planning rules; results of papers II-V are duplicated using this method. Also describes the use to date of proton nuclear magnetic resonance data.

25. Buchanan, B. G., A. M. Duffield, and A. V. Robertson. An application of artificial intelligence to the interpretation of mass spectra. In G. W. A. Milne (Ed.), Mass spectrometry: Techniques and applications. New York: Wiley-Interscience, 1971.

    The most comprehensive introduction to DENDRAL for chemists. Gives a very clear description of the acyclic generator and of the DENDRAL notation. Also describes the early version of the Heuristic DENDRAL program, basically as it appeared in Buchanan, Sutherland, and Feigenbaum (1969). The first part of the paper is good introductory material for understanding the basic concepts of the generator. Presents extensively the results of the DENDRAL program to 1971.

26. Buchanan, B. G., E. A. Feigenbaum, and J. Lederberg. A heuristic programming study of theory formation in science. *Second International Joint Conference on Artificial Intelligence*. London: The British Computer Society, 1971, 40-50. Also *Stanford Artificial Intelligence Project Memo AIM-145*, and *Report No. CS-221*. Stanford, Calif.: Stanford University, Computer Science Department, 1971.

    The first description of the developing Meta-DENDRAL program, presented as the first steps toward a goal of the Heuristic DENDRAL project: studying processes underlying theory formation. A fairly general nontechnical discussion of the Heuristic DENDRAL program as a specific inference maker, whose inference-making processes are to be studied.

27. Buchs, A., A. B. Delfino, C. Djerassi, A. M. Duffield, B. G. Buchanan, E. A. Feigenbaum, J. Lederberg, G. Schroll, and G. L. Sutherland. The application of artificial intelligence in the interpretation of low-resolution mass spectra. In A. Quayle (Ed.), *Advances in mass spectrometry, volume 5*. London: The Institute of Petroleum, 1971, 314-318.

    A brief, somewhat technical paper for the chemistry audience. Describes the DENDRAL approach to a general computer interpretation of low-resolution mass spectra of organic compounds and presents the results obtained with aliphatic amines. The program uses only the Preliminary Inference Maker and the Structure Generator.

28. Feigenbaum, E. A., B. G. Buchanan, and J. Lederberg. On generality and problem solving: A case study using the DENDRAL program. In B. Meltzer and D. Michie (Eds.), *Machine intelligence 6*. Edinburgh: Edinburgh University Press, 1971, 165-190. Also *Stanford Artificial Intelligence Project Memo AIM-131* and *Report No. CS176*. Stanford, Calif.: Stanford University, Computer Science Department, 1970.

    A paper for the artificial intelligence community. A very good summary. Uses the design of Heuristic DENDRAL and its performance on the problems it has solved as a springboard for discussing design for generality, performance problems attendant on too much generality, the coupling of expertise to the general problem-solving processes, and the relationship between generality and expertness. Uses the terminology of Planner-Structure Generator-Predictor. Discusses DENDRAL as a combination of universal and "big switch" approaches to problem solving.

29. Buchanan, B. G., and J. Lederberg. The heuristic DENDRAL program for explaining empirical data. In C. V. Freiman (Ed.), *Information Processing 71, Proceedings of the IFIP Congress 71*. Amsterdam: Elsevier-North Holland, 1972, 1, 179-188. Also *Stanford Artificial Intelligence Project Memo AIM-141* and *Report No. CS-203*. Stanford, Calif.: Stanford University, Computer Science Department, 1971.

    Brief paper written for the computer science audience. Gives the global picture of DENDRAL without going into a detailed presentation. Uses the format of Planner, Structure Generator, and Evaluator. Presents all the DENDRAL results up to 1971.

30. Buchanan, B. G., E. A. Feigenbaum, and N. S. Sridharan. Heuristic theory formation: Data interpretation and rule formation. In B. Meltzer and D. Michie (Eds.), *Machine intelligence 7*. Edinburgh: Edinburgh University Press, 1972, 267-290.

    A Meta-DENDRAL paper for the computer science audience. Describes a program developed to deal with estrogenic steroids, written to test quickly some of the ideas developed on theory formation. Program aimed at discovering theories that will help predict mass spectra for molecules given their chemical structure. Discusses the rule-formation task for Meta-DENDRAL in the context of estrogenic steroids.

31. Lederberg, J. Rapid calculation of molecular formulas from mass values. *Journal of Chemical Education*, 1972, 49, 613.

    A brief technical note addressed to the chemistry education audience. Presents a simple method for the calculation of molecular compositions consistent with a given range of mass values. This compilation is a greatly shortened version of the tables published in Lederberg (1964a).

32. J. Lederberg. Use of a computer to identify unknown compounds: The automation of scientific inference. In G. R. Waller (Ed.), *Biochemical applications of mass spectrometry*. New York: Wiley-Interscience, 1972, 193–207.

    Paper VII in the series of DENDRAL papers for chemists.

33. Smith, D. H., B. G. Buchanan, R. S. Engelmore, A. M. Duffield, A. Yeo, E. A. Feigenbaum, J. Lederberg, and C. Djerassi. Applications of artificial intelligence for chemical inference, VIII. An approach to the computer interpretation of the high resolution mass spectra of complex molecules. Structure elucidation of estrogenic steroids. *Journal of the American Chemical Society*, 1972, 94:17, 5962–5971.

    Paper VIII in the series of DENDRAL papers for chemists. The first description of the PLANNER program, basically in the same form as it exists at present: a general program that is written to permit the chemist to enter information about heuristics in mass spectrometry in a uniform way, as opposed to a detailed presentation of such information, as in the previous version. Describes the application of PLANNER to estrogens, the most extensive use of PLANNER so far. This paper is the basis for Chapter 5 of this book.

34. Buchanan, B. G., and N. S. Sridharan. Analysis of behavior of chemical molecules: Rule formation on non-homogeneous classes of objects. *Third International Joint Conference on Artificial Intelligence*, Menlo Park, Calif.: Stanford Research Institute, Publications Department, 1973, 67–76. Also *Stanford Artificial Intelligence Laboratory Memo AIM-215* and *STAN-CS-73-387*. Stanford, Calif.: Stanford University, Computer Science Department, 1973.

    A description of Meta-DENDRAL for the computer science audience. The program is given mass spectrometry data from several chemical molecules, separates these molecules into subclasses, and selects from the space of all explanatory processes the characteristic processes for each subclass. Discusses some results of the program.

35. Buchanan, B. G. Review of Hubert Dreyfus's What computers can't do: A critique of artificial reason, *Computing Reviews*, 1973, 18–21. Also *Stanford Artificial Intelligence Project Memo AIM-181* and *STAN-CS-72-325*. Stanford, Calif.: Stanford University, Computer Science Department, 1972.

    Review of a controversial critique of artificial intelligence. Takes the position that the philosophical approach of the book is interesting, but the attack on artificial intelligence is not well reasoned.

36. Smith, D. H., B. G. Buchanan, R. S. Engelmore, H. Adlercreutz, and C. Djerassi. Applications of artificial intelligence for chemical inference, IX. Analysis of mixtures without prior separation as illustrated for estrogens. *Journal of the American Chemical Society*, 1973, 95:18, 6078–6084.

    Paper IX in the series of DENDRAL papers for chemists. Describes the use of PLANNER in the analysis of mixtures, as illustrated for estrogens. Essentially the same program as before but now used to analyze mixtures of compounds, which go through the mass spectrometer without prior separation. The program is able to discriminate these, determining from the total composite spectrum which pieces of it correspond to which compounds in the mixture. (The problem of analyzing mixtures is one that chemists find especially difficult.) The program uses a variety of mass spectra, from unseparated mixtures, including high-resolution mass spectra, low-ionizing voltage spectra, and metastable ion spectra.

37. Smith, D. H., B. G. Buchanan, W. C. White, E. A. Feigenbaum, J. Lederberg, and C. Djerassi. Applications of artificial intelligence for chemical inference-X. Intsum. A data interpretation and summary program applied to the collected mass spectra of estrogenic steroids. *Tetrahedron*, 1973, 29, 3117–3134.

Paper X in the series of DENDRAL papers for chemists. A description of INTSUM, one of the Meta-DENDRAL programs. Application of the program to high-resolution mass spectral data from estrogenic steroids. Important because this paper describes the first new chemistry (mass spectrometry) results from Meta-DENDRAL.

38. Carhart, R. E., and C. Djerassi. Applications of artificial intelligence for chemical inference, part XI. Analysis of carbon-13 nuclear magnetic resonance data for structure elucidation of acyclic amines. *Journal of the Chemical Society*, Perkin Transactions II, 1973, 1753–1759.
Paper XI in the series of DENDRAL papers for chemists. Describes the use of a PLANNER with carbon-13 nuclear magnetic resonance data. Describes a computer program called AMINE, which uses a set of predictor rules to deduce the structures of acyclic amines from their empirical formulas and carbon-13 nuclear magnetic resonance spectra. Important extension of DENDRAL ideas beyond mass spectrometry.

39. Sridharan, N. S. Computer generation of vertex-graphs. *STAN-CS-73-381*. Stanford, Calif: Stanford University, Computer Science Department, 1973.
A fairly technical paper directed to the mathematics and computer-programming audience. Describes the cyclic structure generator; deals with the basic set of vertex graphs that the cyclic structure generator draws upon.

40. Sridharan, N. S. Search strategies for the task of organic chemical synthesis. *Stanford Artificial Intelligence Laboratory Memo AIM-217* and *STAN-CS-73-391*. Stanford, Calif.: Stanford University, Computer Science Department, 1973. Presented at the Third International Joint Conference on Artificial Intelligence, Stanford, California, 1973.
A paper for the computer science audience. Describes a computer program, written elsewhere, that successfully discovers syntheses for complex organic molecules. This paper describes the definition of the search space and strategies for heuristic search. This work is of tangential interest to the DENDRAL activities.

41. Sridharan, N. S., H. Gelernter, A. J. Hart, W. F. Fowler, and H. J. Shue. A heuristic program to discover syntheses for complex organic molecules. *Stanford Artificial Intelligence Laboratory Memo AIM-205* and *STAN-CS-73-370*. Stanford, Calif.: Stanford University, Computer Science Department, 1973.
A paper for the computer science audience. Describes a synthesis program, developed elsewhere, for complex organic molecules, giving its organization as a heuristic search and discussing the design of the problem-solving tree and the search procedures. This program is not directly related to the DENDRAL programs.

42. Brown, H., L. Hjelmeland, and L. M. Masinter. Constructive graph labeling using double cosets. *Discrete Mathematics*, 1974, 7, 1–30. Also *STAN-CS-72-318*. Stanford, Calif.: Stanford University, Computer Science Department, 1972.
A highly technical mathematical paper, describing the mathematical basis of the cyclic generator, addressed to graph theorists. Presents algorithms for labeling graphs based on a group theoretic formulation of the labeling problem. Companion to Brown and Masinter (1974).

43. Brown, H., and L. M. Masinter. Algorithm for the construction of the graphs of organic molecules. *Discrete Mathematics*, 8, 1974, 227–244. Also *STAN-CS-73-361*. Stanford, Calif.: Stanford University, Computer Science Department, 1973.
A highly detailed mathematical description of the cyclic generator algorithm, including some theorems to show that it works: for example, that the algorithm produces no redundancies. A technical paper suited for the mathematician or computer programmer. Companion to Brown, Hjelmeland, and Masinter (1974a).

44. Buchanan, B. G. Scientific theory formation by computer. In *Proceedings of NATO Advanced Study Institute on Computer Oriented Learning Processes*, 1974, Bonas, France.
A general tutorial on Meta-DENDRAL notions. Presents artificial intelligence and chemistry information sufficient to understand the functioning of Meta-DENDRAL. Suitable for a general audience.

45. Feigenbaum, E. A. Computer applications: Introductory remarks. In *Proceedings of Federation of American Societies for Experimental Biology*, 1974, 33:12, 2331–2332.
Introductory remarks to a computer science audience on computer applications in artificial intelligence research. Describes the basic concepts on which artificial intelligence research is

based, gives performance levels of existing programs, and gives an introduction to the concept of knowledge-based systems.

46. Masinter, L. M., N. S. Sridharan, J. Lederberg, and D. H. Smith. Applications of artificial intelligence for chemical inference, XII. Exhaustive generation of cyclic and acyclic isomers. *Journal of the American Chemical Society*, 1974, 96:25, 7702–7714. Also *Stanford Artificial Intelligence Laboratory Memo AIM-216* and *STAN-CS-73-389*. Stanford, Calif.: Stanford University, Computer Science Department, 1973.

    Paper XII in the series of DENDRAL papers for chemists. Another description of the cyclic generator; does not require special knowledge of mathematics or computer programming but is a fairly exhaustive and detailed presentation. Chemistry analog of Brown and Masinter (1974).

47. Masinter, L. M., N. S. Sridharan, R. E. Carhart, and D. H. Smith. Applications of artificial intelligence for chemical inference, XIII. Labeling of objects having symmetry. *Journal of the American Chemical Society*, 1974, 96:25, 7714–7723.

    Paper XIII in the series of DENDRAL papers for chemists. Concerns the cyclic generator. The "labeling of objects having symmetry" refers to a part of the cyclic generator. The paper is divided into sections presenting a brief tutorial on the nature of the problem and an introduction to the terminology found in more technical treatment, a textual description of a method for the solution to this type of problem, a summary of the procedure in a more algorithmic form, generalizations on the basis of the algorithm, and a sample application of the method to a complex isomerism problem in organic chemistry. Chemistry analog of Brown, Hjelmeland, and Masinter (1974a) and companion to Masinter et al. (1974b).

48. Michie, D., and B. G. Buchanan. Current status of the heuristic DENDRAL program for applying artificial intelligence to the interpretation of mass spectra. In R. A. G. Carrington (Ed.), *Computers for spectroscopists*. London: Adam Hilger 1974, 114–131. Also *Experimental Programming Report No. 32*, under the title *Artificial intelligence in mass spectroscopy: A review of the heuristic DENDRAL program*. Edinburgh: University of Edinburgh, School of Artificial Intelligence, 1973.

    A technical paper intended for mass spectrometrists. A concise introduction to the DENDRAL programs (in the form of the Planner-Structure Generator-Predictor-Evaluator). Presents an introduction to artificial intelligence work for the chemist.

49. Smith, D. H., L. M. Masinter, and N. S. Sridharan. Heuristic DENDRAL: Analysis of molecular structure. In W. T. Wipke, S. Heller, R. N. Feldman, and E. Hyde (Eds.), *Proceedings of the NATO/CNNA Advanced Study Institute on Computer Representation and Manipulation of Chemical Information*. New York: Wiley, 1974, 287–315.

    A straightforward description of the cyclic generator algorithm in nonmathematical terminology for chemists. (This paper contrasts with Brown, Hjelmeland, and Masinter (1974a), which is a rigorous mathematical description of this algorithm, and with Brown and Masinter (1974), which is also a highly formalized description of the algorithm. In the current paper, the algorithm is described in ordinary language.) Describes the generation, representation, and manipulation of molecular structures in the context of the Heuristic DENDRAL algorithm, with emphasis on the cyclic generator. Also describes the then current work on the Meta-DENDRAL program.

50. Buchanan, B. G. Applications of artificial intelligence to scientific reasoning. In *Proceedings of the Second USA-Japan Computer Conference*. American Federation of Information Processing Societies Press, 1975.

    A paper for computer scientists. Describes the design criteria for heuristic systems. Describes the Heuristic DENDRAL and Meta-DENDRAL programs from the point of view of the design choices that have made the computer programs transferable to the chemistry laboratory.

51. Carhart, R. E., S. M. Johnson, D. H. Smith, B. G. Buchanan, R. G. Dromey, and J. Lederberg. Networking and a collaborative research community: A case study using the DENDRAL programs. In P. Lykos (Ed.), *Proceedings of the American Chemical Society Symposium on Networking and Chemistry*, Chicago, August 1975.

    Describes the SUMEX-AIM computer facility in general and the DENDRAL programs in parti-

cular as examples of shareable programming efforts. Gives a good overview of the working DENDRAL programs and an assessment of the SUMEX-AIM facility.
52. Dromey, R. G., B. G. Buchanan, D. H. Smith, J. Lederberg, and C. Djerassi. Applications of artificial intelligence for chemical inference, XIV. A general method for predicting molecular ions in mass spectra. *Journal of Organic Chemistry*, 1975, **40**:6, 770–774.
Paper XIV in the series of DENDRAL papers for chemists. Describes the MOLION program, discussed in Chapter 5 of this book.
53. Smith, D. H. Applications of artificial intelligence for chemical inference. Constructive graph labeling applied to chemical problems. Chlorinated hydrocarbons. *Analytical Chemistry*, 1975, **47**:7, 1176–1179.
Paper XV in the series of DENDRAL papers for chemists. Describes an application of the labeling algorithm in the cyclic generator to chlorinated hydrocarbons. Application of ideas discussed in Masinter et al. (1974a).
54. Carhart, R. E., D. H. Smith, H. Brown, and N. S. Sridharan. Applications of artificial intelligence for chemical inference, XVI. Computer generation of vertex-graphs and ring systems. *Journal of Chemical Information and Computer Sciences* (formerly *Journal of Chemical Documentation*), 1975, **15**:2, 124–130.
Paper XVI in the series of DENDRAL papers for chemists. Describes a program for enumerating vertex graphs that is used to construct the CATALOG for the cyclic generator.
55. Carhart, R. E., D. H. Smith, H. Brown, and C. Djerassi. Applications of artificial intelligence for chemical inference, XVII. An approach to computer-assisted elucidation of molecular structure. *Journal of the American Chemical Society*, 1975, **97**:20, 5755–5762.
Paper XVII in the series of DENDRAL papers for chemists. Describes the CONGEN program. Another description for chemists of the cyclic generator, mentioning heuristic constraints for the first time, together with a moderately elaborate example of how the program works on a problem that is of potential interest to working chemists.
56. Smith, D. H. The scope of structural isomerism. *Journal of Chemical Information and Computer Sciences*, 1975, **15**:4, 203–207.
Paper XVIII in the series of DENDRAL papers for chemists.
57. Smith, D. H., J. P. Konopelski, and C. Djerassi. Applications of artificial intelligence for chemical inference, XIX. Computer generation of ion structures. *Organic Mass Spectrometry*, 1976, **11**, 86-100.
58. Carhart, R. E., and D. H. Smith. Applications of artificial intelligence for chemical inference, XX. 'Intelligent' use of constraints in computer-assisted structure elucidation. *Computers in Chemistry*, 1976, **1**, 79.
59. Cheer, C., D. H. Smith, C. Djerassi, B. Tursch, J. C. Braekman, and D. Daloze. Applications of artificial intelligence for chemical inference, XXI. The computer-assisted identification of [+]− palustrol in the marine organism *cespitularia ap.*, aff. *subvirdis*. *Tetrahedron*, 1976, **32**, 1807.
60. Buchanan, B. G., D. H. Smith, W. C. White, R. Gritter, E. A. Feigenbaum, J. Lederberg, and C. Djerassi. Applications of artificial intelligence for chemical inference, XXII. Automatic rule formation in mass spectrometry by means of the Meta-DENDRAL program. *Journal of the American Chemical Society*, 1976, **98**, 6168.
61. Dromey, R. G., M. J. Stefik, T. Rindfleisch, and A. M. Duffield. Extraction of mass spectra free of background and neighboring component contributions from gas chromatography/mass spectrometry data. *Analytical Chemistry*, 1976, **48**, 1368.
62. Varkony, T. H., R. E. Carhart, and D. H. Smith. Applications of artificial intelligence for chemical inference, XXIII. Computer-assisted structure elucidation. Modelling chemical reaction sequences used in molecular structure problems. In W. T. Wipke (Ed.), *Computer-assisted organic synthesis*. Washington, D.C.: American Chemical Society, 1977.
63. Smith, D. H., and R. E. Carhart. Applications of artificial intelligence for chemical inference, XXIV. Structural isomerism of mono- and sesquiterpenoid skeletons. *Tetrahedron*, 1976, **32**, 2513.
64. Carhart, R. E. A model-based approach to the teletype printing of chemical structures. *Journal of Chemical Information and Computer Sciences*, 1976, **16**, 82.

65. Dromey, R. G., M. J. Stefik, T. C. Rindfleisch, and A. M. Duffield. Extraction of mass spectra free of background and neighboring component contributions from gas chromatography/mass spectrometry data. *Analytical Chemistry*, 1976, **48**, 1368.
66. Buchanan, B. G., and D. Smith. Computer assisted chemical reasoning. In E. V. Ludena, N. H. Sabelli, and A. C. Wahl (Eds.), *Computers in chemical education and research*. New York: Plenum Press, 1977.
67. Schwenzer, G. E., and T. M. Mitchell. Computer assisted structure elucidation using automatically acquired 13C NMR rules. In D. Smith (Ed.), *Computer assisted structure elucidation*, ACS Symposium Series, 1977, **54**, 58.
68. Nourse, J. G. Generalized stereoisomerization modes. *Journal of the American Chemical Society*, 1977, **99**, 2063.
69. Varkony, T. H., R. E. Carhart, and D. H. Smith. Computer assisted structure elucidation, ranking of candidate structures, based on comparison between predicted and observed mass spectra. In *Proceedings of the Twenty-Fifth Annual Conference on Mass Spectrometry and Allied Topics*, Washington, D. C., 1977.
70. Mitchell, T. M. Version spaces: A candidate elimination approach to rule learning. *Proceedings of the Fifth IJCAI*, **1**, 305, August 1977.
71. Feigenbaum, E. A., J. Lederberg, and B. G. Buchanan. Final technical report for contract period August 1, 1973 through July 31, 1977, *Contract DAHC-73-C-0435*. Submitted to Advanced Research Projects Agency, Department of Defense, October, 1977.
72. Mitchell, T. M., and G. M. Schwenzer. Applications of artificial intelligence for chemical inference, XXV. A computer program for automated empirical 13C NMR rule formation. *Organic Magnetic Resonance*, 1978, **11**, No. 8, 378.
73. Schwenzer, G. E. Applications of artificial intelligence for chemical inference, XXVI. Analysis of C-13 NMR for mono-hydroxy steroids incorporating geometric distortions. *Journal of Organic Chemistry*, 1978, **43**, No. 6.
74. Buchanan, B. G., T. Mitchell, R. G. Smith, and C. R. Johnson, Jr. Models of learning systems. In J. Belzer (Ed.), *Encyclopedia of computer sciences and technology*, **11**. New York: Marcel Dekker, 1978.
75. Buchanan, B. G., and E. A. Feigenbaum. DENDRAL and Meta-DENDRAL: Their applications dimension. *Artificial Intelligence*, 1978, **11**.
76. Varkony, T. H., R. E. Carhart, D. H. Smith, and C. Djerassi. Computer-assisted simulation of chemical reaction sequences: Applications to problems of structure elucidation. *Journal of Chemical Information and Computer Sciences*, 1978, **18**, 168.
77. Carhart, R. E. Erroneous claims concerning the perception of topological symmetry. *Journal of Chemical Information and Computer Sciences*, 1978, **18**, 197.
78. Gray, N. A. B., D. H. Smith, T. H. Varkony, R. E. Carhart, and B. G. Buchanan. Use of a computer to identify unknown compounds: The automation of scientific inference. In G. R. Waller (Ed.), *Biochemical applications of mass spectrometry (supplement)*. New York: Wiley-Interscience, 1978.
79. Smith, D. H., and P. C. Jurs. Prediction of 13C NMR chemical shifts. *Journal of the American Chemical Society*, 1978, **100**, 3316.
80. Smith, D. H., T. C. Rindfleisch, and W. J. Yeager. Exchange of comments: Analysis of complex volatile mixtures by a combined gas chromatography–mass spectrometry system. *Analytical Chemistry*, 1978, **50**, 1585.
81. Nourse, J. G. Applications of the permutation group in dynamic stereochemistry. In *The permutation group in physics and chemistry. Lecture notes in chemistry, volume 12*. New York: Springer-Verlag, 1979, 28.
82. Nourse, J. G. Application of the permutation group to stereoisomer generation for computer assisted structure elucidation. In *The permutation group in physics and chemistry. Lecture notes in chemistry, volume 12*. New York: Springer-Verlag, 1979, 19.
83. Nourse, J. G., Applications of artificial intelligence to chemical inference, 28. The configuration symmetry group and its application to stereoisomer generation, specification and enumeration. *Journal of the American Chemical Society*, 1979, **101**, 1210.

84. Smith, D. H., and R. E. Carhart. Structure elucidation based on computer analysis of high and low resolution mass spectral data. In M. L. Gross (Ed.), *Proceedings of the Symposium on Chemical Applications of High Performance Spectrometry.* Washington, D. C.: American Chemical Society, 1978, 325.
85. Varkony, T., D. Smith and C. Djerassi. Computer-assisted structure manipulation: Studies in the biosynthesis of natural products. *Tetrahedron*, 1978, **34**, 841-852.
86. Mitchell, T. M. Version spaces: An approach to concept learning. Doctoral dissertation. Stanford University, Department of Electrical Engineering, December 1978.
87. Wegmann, A. Variations in mass spectral fragmentation produced by active sites in a mass spectrometer source. *Analytical Chemistry*, 1978, **50**, No. 6, 830-832.
88. Nourse, J. G., R. E. Carhart, D. H. Smith, and C. Djerassi. Applications of artificial intelligence to chemical inference, 29. Exhaustive generation of stereoisomers for structure elucidation. *Journal of the American Chemical Society*, 1979, **101**, 1216.
89. Fitch, W. L., P. J. Anderson, and D. H. Smith. Isolation, identification and quantitation of urinary organic acids. *Journal of Chromatography*, 1979, **162**, 249-259.
90. Fitch, W. L., E. T. Everhart, and D. H. Smith. Characterization of carbon black adsorbates and artifacts formed during extraction. *Analytical Chemistry*, 1978, **50**, 2122-2126.
91. Fitch, W. L., and D. H. Smith. Analysis of adsorption properties and adsorbed species of commercial polymeric carbons. *Environmental Science and Technology*, 1979, **13**, 341-346.
92. Djerassi, C., D. H. Smith, and T. H. Varkony. A novel role of computers in the natural products field. *Naturwissenschaften*, 1979, **66**, 9.
93. Rindfleisch, T. C., and D. H. Smith. Advances in data acquisition and analysis systems for applications of gas chromatography/mass spectrometry. In G. R. Waller (Ed.), *Biomedical applications of mass spectrometry.* New York: Wiley-Interscience, 1978.
94. Buchanan, B. G., and T. Mitchell. Model-directed learning of production rules. In D. A. Waterman and F. Hayes-Roth (Eds.), *Pattern directed inference systems.* New York: Academic Press, 1978.
95. Buchanan, B. G. Issues of representation in conveying the scope and limitations of intelligent assistant programs. In J. E. Hayes, D. Michie, and L. I. Mikulich, *Machine intelligence 9.* Chichester, England: Ellis Horwood Ltd. and New York: Wiley, 1979.
96. Nourse, J. G., and D. H. Smith. Nonnumerical mathematical methods in the problem of stereoisomer generation. *Match*, 1979, **6**, 259.

REFERENCES

Beynon, J. H., and R. M. Caprioli. Metastable ions as an aid in the interpretation of mass spectra. In G. R. Waller (Ed.), *Biochemical applications of mass spectrometry.* New York: Wiley-Interscience, 1972, 157-176.

Boden, M. *Artificial intelligence and natural man.* New York: Basic Books, 1977.

Bronowski, J. The logic of the mind. *American Scientist*, 1966, **54**, 1-14.

Brown, H., L. Hjelmeland, and L. M. Masinter. Constructive graph labeling using double cosets. *Discrete Mathematics*, 1974, 7, 1-30. Also *STAN-CS-72-318.* Stanford, Calif.: Stanford University, Computer Science Department, 1972.

Brown, H., and L. M. Masinter. Algorithm for the construction of the graphs of organic molecules. *Discrete Mathematics*, 1974, 8, 227-244, Also *STAN-CS-73-361.* Stanford Calif.: Stanford University, Computer Science Department, 1973.

Buchanan, B. G. Review of Hubert Dreyfus's What computers can't do: A critique of artificial reason, *Computing Reviews*, 1973, 18-21. Also *Stanford Artificial Intelligence Project Memo AIM-181* and *STAN-CS-72-325.* Stanford, Calif.: Stanford University, Computer Science Department, 1972.

Buchanan, B. G. Applications of artificial intelligence to scientific reasoning. In *Proceedings of the Second USA-Japan Computer Conference*. American Federation of Information Processing Societies Press, 1975.

Buchanan, B. G. Issues of representation in conveying the scope and limitations of intelligent assistant programs. In J. E. Hayes, D. Michie, and L. I. Mikulich, *Machine intelligence 9*. Chichester, England: Ellis Horwood Ltd. and New York: Wiley, 1979.

Buchanan, B. G., A. M. Duffield, and A. V. Robertson. An application of artificial intelligence to the interpretation of mass spectra. In G. W. A. Milne (Ed.), *Mass spectrometry: Techniques and applications*. New York: Wiley-Interscience, 1971.

Buchanan, B. G., and E. A. Feigenbaum. DENDRAL and meta-DENDRAL: Their applications dimension. *Artificial Intelligence*, 1978, **11**.

Buchanan, B.G., E. A. Feigenbaum, and J. Lederberg. A heuristic programming study of theory formation in science. *Second International Joint Conference on Artificial Intelligence*. London: The British Computer Society, 1971, 40–50. Also *Stanford Artificial Intelligence Project Memo AIM-145* and *Report No. CS-221*. Stanford, Calif.: Stanford University, Computer Science Department, 1971.

Buchanan, B. G., E. A. Feigenbaum, and N. S. Sridharan. Heuristic theory formation: Data interpretation and rule formation. In B. Meltzer and D. Michie (Eds.), *Machine intelligence 7*. Edinburgh: Edinburgh University Press, 1972, 267–290.

Buchanan, B. G., and J. Lederberg, The heuristic DENDRAL program for explaining empirical data. In C. V. Freiman (Ed.), *Information Processing 71, Proceedings of the IFIP Congress 71*. Amsterdam: Elsevier-North Holland, 1972, **1**, 179–188. Also *Stanford Artificial Intelligence Project Memo AIM-141* and *Report No. CS-203*. Stanford, Calif.: Stanford University, Computer Science Department, 1971.

Buchanan, B. G., and T. Mitchell. Model-directed learning of production rules. In D. A. Waterman and F. Hayes-Roth (Eds.), *Pattern directed inference systems*. New York: Academic Press, 1978.

Buchanan, B. G., T. Mitchell, R. G. Smith, and C. R. Johnson, Jr. Models of learning systems. In J. Belzer (Ed.), *Encyclopedia of computer sciences and technology*, *11*, New York: Marcel Dekker, 1978.

Buchanan, B. G., and D. H. Smith. Computer assisted chemical reasoning. In E. V. Ludena, N. H. Sabelli, and A. C. Wahl (Eds.), *Computers in chemical education and research*. New York: Plenum Press, 1977.

Buchanan, B. G., D. H. Smith, W. C. White, R. Gritter, E. A. Feigenbuam, J. Lederberg, and C. Djerassi. Applications of artificial intelligence for chemical inference, XXII. Automatic rule formation in mass spectrometry by means of the meta-DENDRAL program. *Journal of the American Chemical Society*, 1976.

Buchanan, B. G., and N. S. Sridharan. Analysis of behavior of chemical molecules: Rule formation on non-homogeneous classes of objects. *Third International Joint Conference on Artificial Intelligence*. Menlo Park, Calif.: Stanford Research Institute, Publications Department, 1973, 67–76. Also *Stanford Artificial Intelligence Laboratory Memo AIM-215* and *STAN-CS-73-387*. Stanford, Calif.: Stanford University, Computer Science Department, 1973.

Buchanan, B. G., G. L. Sutherland, and E. A. Feigenbaum. Heuristic DENDRAL: A program for generating explanatory hypotheses in organic chemistry. In B. Meltzer and D. Michie (Eds.), *Machine intelligence 4*. Edinburgh: Edinburgh University Press, 1969, 209–254. Also *Stanford Artificial Intelligence Project Memo AI-62*. Stanford, Calif.: Stanford University, Computer Science Department, 1968.

Buchanan, B. G., G. L. Sutherland, and E. A. Feigenbaum. Rediscovering some problems of artificial intelligence in the context of organic chemistry. In B. Meltzer and D. Michie (Eds.), *Machine intelligence 5*. Edinburgh: Edinburgh University Press, 1970, 253–280. Also *Stanford Artificial Intelligence Project Memo AIM-99*, under the title "Toward an understanding of information processes of scientific inference in the context of organic chemistry." Stanford, Calif.: Stanford University, Computer Science Department, 1969.

Buchs, A., A. B. Delfino, C. Djerassi, A. M. Duffield, B. G. Buchanan, E. A. Feigenbaum, J. Lederberg, G. Schroll, and G. L. Sutherland. The application of artificial intelligence in the interpretation of low-resolution mass spectra. In A. Quayle (Ed.), *Advances in mass spectrometry*, volume 5. London: The Institute of Petroleum, 1971, 314-318.

Buchs, A., A. B. Delfino, A. M. Duffield, C. Djerassi, B. G. Buchanan, E. A. Feigenbaum, and J. Lederberg. Applications of "artificial intelligence" for chemical inference, VI. Approach to a general method of interpreting low resolution mass spectra with a computer. *Helvetica Chimica Acta*, 1970(a), **53**:6, 1394-1417.

Buchs, A., A. M. Duffield, G. Schroll, C. Djerassi, A. B. Delfino, B. G. Buchanan, G. L. Sutherland, E. A. Feigenbaum, and J. Lederberg. Applications of artificial intelligence for chemical inference, IV. Saturated amines diagnosed by their low resolution mass spectra and nuclear magnetic resonance spectra. *Journal of the American Chemical Society*, 1970(b), **92**:23, 6831-6838.

Carhart, R. E. A model-based approach to the teletype printing of chemical structures. *Journal of Chemical Information and Computer Sciences*, 1976, **16**, 82.

Carhart, R. E. Erroneous claims concerning the perception of topological symmetry. *Journal of Chemical Information and Computer Sciences*, 1978, **18**, 197.

Carhart, R. E., and C. Djerassi. Applications of artificial intelligence for chemical inference, part XI. Analysis of carbon-13 nuclear magnetic resonance data for structure elucidation of acyclic amines. *Journal of the Chemical Society*, Perkin Transactions II, 1973, 1753-1759.

Carhart, R. E., S. M. Johnson, D. H. Smith, B. G. Buchanan, R. G. Dromey, and J. Lederberg. Networking and a collaborative research community: A case study using the DENDRAL programs. In P. Lykos (Ed.), *Proceedings of the American Chemical Society Symposium on Networking and Chemistry*, Chicago, August 1975(a).

Carhart, R. E., and D. H. Smith. Applications of artificial intelligence for chemical inference, XX. 'Intelligent' use of constraints in computer-assisted structure elucidation. *Computers in Chemistry*, 1976, **1**, 79.

Carhart, R. E., D. H. Smith, H. Brown, and C. Djerassi. Applications of artificial intelligence for chemical inference, XVII. An approach to computer-assisted elucidation of molecular structure. *Journal of the American Chemical Society*, 1975(b), **97**:20, 5755-5762.

Carhart, R. E., D. H. Smith, H. Brown, and N. S. Sridharan. Applications of artificial intelligence for chemical inference, XVI. Computer generation of vertex-graphs and ring systems. *Journal of Chemical Information and Computer Sciences*, 1975(c), **15**:2, 124-130.

Cheer, C., D. H. Smith, C. Djerassi, B. Tursch, J. C. Braekman, and D. Daloze. Applications of artificial intelligence for chemical inference, XXI. The computer-assisted identification of [+]-palustrol in the marine organism *cespitularia ap.*, aff. *subvirdis*. *Tetrahedron*, 1976, **32**, 1807.

Churchman, C. W., and B. G. Buchanan. On the design of inductive systems: Some philosophical problems. *British Journal for the Philosophy of Science*, 1969, **20**, 311-323.

Davis, R. Applications of meta level knowledge to the construction, maintenance and use of large knowledge bases. Ph.D. dissertation, Department of Computer Science, Stanford University, July 1976.

Djerassi, C., D. H. Smith, and T. H. Varkony. A novel role of computers in the natural products field. *Naturwissenschaften*, 1979, **66**, 9.

Dromey, R. G., B. G. Buchanan, D. H. Smith, J. Lederberg, and C. Djerassi. Applications of artificial intelligence for chemical inference, XIV. A general method for predicting molecular ions in mass spectra. *Journal of Organic Chemistry*, 1975, **40**:6, 770-774.

Dromey, R. G., M. J. Stefik, T. C. Rindfleisch, and A. M. Duffield. Extraction of mass spectra free of background and neighboring component contributions from gas chromatography/mass spectrometry data. *Analytical Chemistry*, 1976, **48**, 1368.

Duffield, A. M., A. V. Robertson, C. Djerassi, B. G. Buchanan, G. L. Sutherland, E. A. Feigenbaum, and J. Lederberg. Applications of artificial intelligence for chemical inference, II. Interpretation of low-resolution mass spectra of ketones. *Journal of the American Chemical Society*, 1969, **91**:11, 2977-2981.

Erman, L. D., and V. R. Lesser. The HEARSAY-II speech understanding system: A tutorial introduction and retrospective view. CMU-CS-78-117. Pittsburgh, Pa.: Department of Computer Science, Carnegie-Mellon University, May 1978.

Ernst, G. W., and A. Newell. *GPS: A case study in generality and problem solving*. New York: Academic Press, 1969.

Friedberg, R. M. A learning machine: part I. *IBM Journal of Research and Development*, 1958, **2**, 2-13.

Feigenbaum, E. A. Artificial intelligence: Themes in the second decade. In A. J. H. Morrell (Ed.), *Information Processing 68, Proceedings of IFIP Congress 1968*. Amsterdam: North-Holland, 1969, volume II, 1008-1023. Also *Stanford Artificial Intelligence Project Memo AI-67*. Stanford, Calif.: Stanford University, Computer Science Department, 1968.

Feigenbaum, E. A. Computer applications: Introductory remarks. In *Proceedings of Federation of American Societies for Experimental Biology*, 1974, **33**:12, 2331-2332.

Feigenbaum, E. A., B. G. Buchanan, and J. Lederberg. On generality and problem solving: A case study using the DENDRAL program. In B. Meltzer and D. Michie (Eds.), *Machine intelligence 6*. Edinburgh: Edinburgh University Press, 1971, 165-190. Also *Stanford Artificial Intelligence Project Memo AIM-131* and *Report No. CS176*. Stanford, Calif.: Stanford University, Computer Science Department, 1970.

Feigenbaum, E. A., J. Lederberg, and B. G. Buchanan. Heuristic DENDRAL: A program for generating explanatory hypotheses in organic chemistry. In B. K. Kinariwala and F. F. Kuo (Eds.), *Proceedings of the Hawaii International Conference on System Sciences*. Honolulu: University of Hawaii Press, 1968, 482-485.

Feigenbaum, E. A., J. Lederberg, and B. G. Buchanan. Final Technical Report for Contract Period August 1, 1973 through July 31, 1977, Contract DAHC-73-C-0435. Submitted to Advanced Research Projects Agency, Department of Defense, October 1977.

Fitch, W. L., P. J. Anderson, and D. H. Smith. Isolation, identification and quantitation of urinary organic acids, *Journal of Chromatography*, 1979, **162**, 249-259.

Fitch, W. L., E. T. Everhart, and D. H. Smith. Characterization of carbon black adsorbates and artifacts formed during extraction. *Analytical Chemistry*, 1978, **50**, 2122-2126.

Fitch, W. L., and D. H. Smith. Analysis of adsorption properties and adsorbed species of commercial polymeric carbons. *Environmental Science and Technology* 1979, **13**, 341-346.

Gaschnig, J. Exactly how good are heuristics?: Toward a realistic predictive theory of best-first search. *Fifth International Joint Conference of Artificial Intelligence-1977*. Proceedings of the conference held August 22-25, 1977 at Massachusetts Institute of Technology, Cambridge, Mass. 434-441.

Gray, N. A. B., D. H. Smith, T. H. Varkony, R. E. Carhart, and B. G. Buchanan. Use of a Computer to identify unknown compounds: The automation of scientific inference. In G. R. Waller (Ed.), *Biochemical applications of mass spectrometry (supplement)*. New York: Wiley-Interscience, 1978.

Green, C. Theorem-proving by resolution as a basis for question-answering. *Machine intelligence 4*. New York: American Elsevier, 1969, 183-205.

Hearn, A. C. REDUCE 2: A system and language for algebraic manipulation. In S. R. Petrick (Ed.), *Proceedings of the ACM Second Symposium on Symbolic and Algebraic Manipulation*. Los Angeles, 1971.

Heller, S. R., G. W. A. Milne, and R. J. Feldman. A computer-based chemical information system. *Science*, 1977, **195**, 4275.

Jennings, K. R. Some aspects of metastable transitions. In G. W. A. Milne (Ed.), *Mass spectrometry: Techniques and applications*. New York: Wiley-Interscience, 1971, 419-458.

Kuhn, T. S. *The structure of scientific revolutions*. Chicago: The University of Chicago Press, 1962.

Laudan, L. Peirce and the trivialization of the self-correcting thesis. In R. N. Giere and R. S. Westfall (Eds.), *Foundations of scientific method: The nineteenth century*. Bloomington: Indiana University Press, 1973.

Lederberg, J. *Computation of molecular formulas for mass spectrometry*. San Francisco: Holden-Day, 1964(a).

Lederberg, J. DENDRAL-64, a system for computer construction, enumeration and notation of organic molecules as tree structures and cyclic graphs, part I. Notational algorithm for tree structures. *Report No. CR-57029* and *STAR No. N65-13158*. National Aeronautics and Space Administration, 1964(b).

Lederberg, J. DENDRAL-64, a system for computer construction, enumeration and notation of organic molecules as tree structures and cyclic graphs, part II. Topology of cyclic graphs. *Report No. CR-68898* and *STAR No. N66-14074*. National Aeronautics and Space Administration, 1965(a).

Lederberg, J. Topological mapping of organic molecules. *Proceedings of the National Academy of Sciences*, 1965(b), 53:1, 134-139.

Lederberg, J. Systematics of organic molecules, graph topology and Hamilton circuits, a general outline of the DENDRAL system. *Report No. CR-68899* and *STAR No. N66-14075*. National Aeronautics and Space Administration, 1966.

Lederberg, J. Hamilton circuits of convex trivalent polyhedra (up to 18 vertices). *American Mathematical Monthly*, 1967, 74:5, 522-527.

Lederberg, J. On line computation of molecular formulas from mass number. *Report No. CR-95977*. National Aeronautics and Space Administration, 1968.

Lederberg, J. Topology of molecules. In the National Research Council's Committee on Support of Research in the Mathematical Sciences (COSRIMS), *The mathematical sciences—A collection of essays*. Published for the National Academy of Sciences (National Research Council). Cambridge: M.I.T. Press, 1969, 37-51.

Lederberg, J. DENDRAL, a system for computer construction, enumeration and notation of organic molecules as tree structures and cyclic graphs, part III. Complete chemical graphs; embedding rings in trees. Technical report, National Aeronautics and Space Administration, 1970.

Lederberg, J. Rapid calculation of molecular formulas from mass values. *Journal of Chemical Education*, 1972(a), 49, 613.

Lederberg, J. Use of a computer to identify unknown compounds: The automation of scientific inference. In G. R. Waller (Ed.), *Biochemical applications of mass spectrometry*. New York: Wiley-Interscience, 1972(b), 193-207.

Lederberg, J., and E. A. Feigenbaum. Mechanization of inductive inference in organic chemistry. In B. Kleinmuntz (Ed.), *Formal representation of human judgment*. New York: Wiley, 1968, 187-218. Also Stanford Artificial Intelligence Project Memo No. 54. Stanford, Calif.: Stanford University, Computer Science Department, 1967.

Lederberg, J., G. L. Sutherland, B. G. Buchanan, and E. A. Feigenbaum. A heuristic program for solving a scientific inference problem: Summary of motivation and implementation. *Stanford Artificial Intelligence Project Memo AIM-104*. Stanford, Calif.: Stanford University, Computer Science Department, 1969(a). Also in R. Banerji and M. D. Mesarovic (Eds.), *Theoretical approaches to nonnumerical problem solving*. New York: Springer-Verlag, 1970.

Lederberg, J., G. L. Sutherland, B. G. Buchanan, E. A. Feigenbaum, A. V. Robertson, A. M. Duffield, and C. Djerassi. Applications of artificial intelligence for chemical inference, I. The number of possible organic compounds. Acyclic structures containing C, H, O, and N. *Journal of the American Chemical Society*, 1969(b), 91:11, 2973-2976.

Lindsay, R. K. *Toward the development of machines which comprehend*. Doctoral dissertation, Carnegie Institute of Technology, 1961.

Lindsay, R. K. In defense of ad hoc systems. In R. C. Schank and K. M. Colby (Eds.), *Computer models of thought and language*. San Francisco: W. H. Freeman, 1973, 372-395.

Martin, W. A., and R. J. Fateman. The MACSYMA system. In S. R. Petrick (Ed.), *Proceedings of the ACM Second Symposium on Symbolic and Algebraic Manipulation*. Los Angeles, 1971.

Masinter, L. M., N. S. Sridharan, R. E. Carhart, and D. H. Smith. Applications of artificial intelligence for chemical inference, XIII. Labeling of objects having symmetry. *Journal of the American Chemical Society*, 1974(a), 96:25, 7714-7723.

Masinter, L. M., N. S. Sridharan, J. Lederberg, and D. H. Smith. Applications of artificial intelligence for chemical inference, XII. Exhaustive generation of cyclic and acyclic isomers. *Journal of the American Chemical Society*, 1974(b), 96:25, 7702-7714. Also Stanford Artificial Intelligence Laboratory Memo AIM-216 and STAN-CS-73-389. Stanford, Calif.: Stanford University, Computer Science Department, 1973.

McCarthy, J. Recursive functions of symbolic expressions and their computation by machine, part 1. *Communications of the ACM*, 1960, 3:4, 184-195.

McLafferty, F. W. *Interpretation of mass spectra: An introduction.* New York: W. A. Benjamin, 1967.
McLafferty, F. W., and R. Venkataraghaven. Computer applications in mass spectometry. In M. L. Gross (Ed.), *High performance mass spectrometry: Chemical applications, ACS symposium series 70,* Washington, D.C.: American Chemical Society, 1978.
Medawar, P. B. *Induction and intuition in scientific thought,* volume 75 in the series *Memoirs of the American Philosophical Society.* Philadelphia: American Philosophical Society, 1969.
Michie, D., and B. G. Buchanan. Current status of the heuristic DENDRAL program for applying artificial intelligence to the interpretation of mass spectra. In R. A. G. Carrington (Ed.), *Computers for spectroscopists.* London: Adam Hilger, 1974, 114-131. Also *Experimental Programming Report No. 32,* under the title *Artificial intelligence in mass spectroscopy: A review of the heuristic DENDRAL program.* Edinburgh: University of Edinburgh, School of Artificial Intelligence, 1973.
Milne, G. W. A. (Ed.). *Mass spectrometry: Techniques and applications.* New York: Wiley-Interscience, 1971.
Mitchell, T. M. Version spaces: A candidate elimination approach to rule learning. *Proceedings of the Fifth IJCAI,* August 1977, **1**, 305.
Mitchell, T. M. Version spaces: An approach to concept learning. Doctoral dissertation, Stanford University, Department of Electrical Engineering, December 1978.
Mitchell, T. M., and G. M. Schwenzer. Applications of artificial intelligence for chemical inference, XXV. A computer program for automated empirical 13C NMR rule formation. *Organic Magnetic Resonance,* 1978, **11**, No. 8, 378.
Mortimer, C. E. *Chemistry: A conceptual approach* (2d ed.). New York: Van Nostrand Reinhold, 1971.
Newell, A. Production systems: Models of control structures. In W. Chase (Ed.), *Visual information processing.* New York: Academic Press, 1972.
Newell, A., and H. A. Simon. *Human problem solving.* Englewood Cliffs, N. J.: Prentice-Hall, 1972.
Nilsson, N. *Problem-solving methods in artificial intelligence.* New York: McGraw-Hill, 1971.
Nourse, J. G. Generalized stereoisomerization modes. *Journal of the American Chemical Society,* 1977, **99**, 2063.
Nourse, J. G. Applications of the permutation group in dynamic sterochemistry. In *The permutation group in physics and chemistry. Lecture notes in chemistry, vol. 12.* New York: Springer-Verlag, 1979(a), 28.
Nourse, J. G. Application of the permutation group to stereoisomer generation for computer assisted structure elucidation. In *The permutation group in physics and chemistry. Lecture notes in chemistry, vol. 12.* New York: Springer-Verlag, 1979(b), 19.
Nourse, J. G. Applications of artificial intelligence to chemical inference, 28. The configuration symmetry group and its application to stereoisomer generation, specification and enumeration. In *Journal of the American Chemical Society,* 1979(c), **101**, 1210.
Nourse, J. G., and D. H. Smith. Nonnumerical mathematical methods in the problem of stereoisomer generation. *Match,* 1979(d), **6**, 259.
Nourse, J. G., R. E. Carhart, D. H. Smith, and C. Djerassi. Applications of artificial intelligence to chemical inference, 29. Exhaustive generation of steroisomers for structure elucidation. *Journal of the American Chemical Society,* 1979(b), **101**, 1216.
Polya, G. *Mathematics and plausible reasoning, volume 2.* Princeton, N. J.: Princeton University Press, 1954.
Popper, K. R. *The logic of scientific discovery.* New York: Harper & Row, 1968.
Raphael, B. *Robot research at Stanford research institute.* Artificial Intelligence Center Technical Note 64, SRI Report 1530. Menlo Park, Calif.: Stanford Research Institute, February 1972.
Reddy, D. R., L. D. Erman, and R. B. Neely. A model and a system for machine recognition of speech. *IEEE Transactions on Audio and Electroacoustics,* 1973, AU-21:3.
Rindfleisch, T. C., and D. H. Smith. Advances in data acquisition and analysis systems for applications of gas chromatography/mass spectrometry. In G. R. Waller (Ed.), *Biomedical applications of mass spectrometry.* New York: Wiley-Interscience, 1978.
Robinson, J. A. A machine-oriented logic based on the resolution principle. *Journal of the ACM,* 1965, **12**:1, 23-41.

Rouvray, D. H. *Chemistry in Britain*, 1974, **10**:11.

Schroll, G., A. M. Duffield, C. Djerassi, B. G. Buchanan, G. L. Sutherland, E. A. Feigenbaum, and J. Lederberg. Applications of artificial intelligence for chemical inference, III. Aliphatic ethers diagnosed by their low-resolution mass spectra and nuclear magnetic resonance data. *Journal of the American Chemical Society*, 1969, **91**:26, 7440–7445.

Schwenzer, G. E. Applications of artificial intelligence for chemical inference, XXVI. Analysis of C-13 NMR for mono-hydroxy steroids incorporating geometric distortions. *Journal of Organic Chemistry*, 1978, **43**, No. 6.

Schwenzer, G. E., and T. M. Mitchell. Computer assisted structure elucidation using automatically acquired 13C NMR rules. In D. Smith (Ed.), *Computer assisted structure elucidation*. ACS Symposium Series, 1977, **54**, 58.

Sheikh, Y. M., A. Buchs, A. B. Delfino, G. Schroll, A. M. Duffield, C. Djerassi, B. G. Buchanan, G. L. Sutherland, E. A. Feigenbaum, and J. Lederberg. Applications of artificial intelligence for chemical inference. An approach to the computer generation of cyclic structures. Differentiation between all the possible isomeric ketones of composition $C_6H_{10}O$. *Organic Mass Spectrometry*, 1970, **4**, 493–501.

Shortliffe, E. H. *Computer-based medical consultations: MYCIN*. New York: American Elsevier, 1976.

Simon, H. A. *The sciences of the artificial*. Cambridge: M. I. T. Press, 1969.

Simon, H. A., and M. Barenfeld. Information-processing analysis of perceptual processes in problem solving. *Psychological Review*, 1969, **76**:5, 473–483.

Smith, D. H. Applications of artificial intelligence for chemical inference. Constructive graph labeling applied to chemical problems. Chlorinated hydrocarbons. *Analytical Chemistry*, 1975(a), **47**:7, 1176–1179.

Smith, D. H. The scope of structural isomerism. *Journal of Chemical Information and Computer Sciences*, 1975(b), **15**:4, 203–207.

Smith, D. H., B. G. Buchanan, R. S. Engelmore, A. M. Duffield, A. Yeo, E. A. Feigenbaum, J. Lederberg, and C. Djerassi. Applications of artificial intelligence for chemical inference, VIII. An approach to the computer interpretation of the high resolution mass spectra of complex molecules. Structure elucidation of estrogenic steroids. *Journal of the American Chemical Society*, 1972, **94**:17, 5962–5971.

Smith, D. H., B. G. Buchanan, R. S. Engelmore, H. Adlercreutz, and C. Djerassi. Applications of artificial intelligence for chemical inference, IX. Analysis of mixtures without prior separation as illustrated for estrogens. *Journal of the American Chemical Society*, 1973(a), **95**:18, 6078–6084.

Smith, D. H., B. G. Buchanan, W. C. White, E. A. Feigenbaum, J. Lederberg, and C. Djerassi. Applications of artificial intelligence for chemical inference-X. Intsum. A data interpretation and summary program applied to the collected mass spectra of estrogenic steroids. *Tetrahedron*, 1973(b), **29**, 3117–3134.

Smith, D. H., and R. E. Carhart. Applications of artificial intelligence for chemical inference, XXIV. Structural isomerism of mono-and sesquiterpenoid skeletons. *Tetrahedron*, 1976, **32**, 2513.

Smith, D. H., and R. E. Carhart. Structure elucidation based on computer analysis of high and low resolution mass spectral data. In M. L. Gross (Ed.), *Proceedings of the Symposium on Chemical Applications of High Performance Spectrometry*. Washington, D. C.: American Chemical Society, 1978, 325.

Smith, D. H., and P. C. Jurs. Prediction of 13C NMR chemical shifts. *Journal of the American Chemical Society*, 1978, **100**, 3316.

Smith, D. H., J. P. Konopelski, and C. Djerassi. Applications of artificial intelligence for chemical inference, XIX. Computer generation of ion structures. *Organic Mass Spectrometry*, 1976, **11**, 86–100.

Smith, D. H., L. M. Masinter, and N. S. Sridharan. Heuristic DENDRAL: Analysis of molecular structure. In W. T. Wipke, S. Heller, R. N. Feldman, and E. Hyde (Eds.), *Proceedings of the NATO/CNNA Advanced Study Institute on Computer Representation and Manipulation of Chemical Information*. New York: Wiley, 1974, 287–315.

Smith, D. H., T. C. Rindfleisch, and W. J. Yeager. Exchange of comments: Analysis of complex volatile mixtures by a combined gas chromatography-mass spectrometry system. *Analytical Chemistry*, 1978, 50, 1585.

Sridharan, N. S. Computer generation of vertex-graphs. *STAN-CS-73-381*. Stanford, Calif.: Stanford University, Computer Science Department, 1973(a).

Sridharan, N. S. Search strategies for the task of organic chemical synthesis. *Stanford Artificial Intelligence Laboratory Memo AIM-217* and *STAN-CS-73-391*. Stanford, Calif.: Stanford University, Computer Science Department, 1973(b). Presented at the Third International Joint Conference on Artificial Intelligence, Stanford, California, 1973.

Sridharan, N. S., H. Gelernter, A. J. Hart, W. F. Fowler, and H. S. Shue. A heuristic program to discover syntheses for complex organic molecules. *Stanford Artificial Intelligence Laboratory Memo AIM-205* and *STAN-CS-73-370*. Stanford, Calif.: Stanford University, Computer Science Department, 1973.

Sutherland, G. L. DENDRAL—A computer program for generating and filtering chemical structures. *Stanford Artificial Intelligence Project Memo No. 49*. Stanford, Calif.: Stanford University, Computer Science Department, 1967.

Sutherland, G. L. Heuristic DENDRAL: A family of LISP programs. *Stanford Artificial Intelligence Project Memo AI-80*. Stanford, Calif.: Stanford University, Computer Science Department, 1969.

Teitelman, W. *INTERLISP reference manual*, 2d ed. Palo Alto, Calif.: Xerox Palo Alto Research Center, 1975.

Toulmin, S. E. *Foresight and understanding*. Bloomington: Indiana University Press, 1961.

VanLehn, K. A. *SAIL User Manual. Stanford Artificial Intelligence Laboratory Memo AIM-204* and *Computer Science Department Report STAN-CS-73-373*. Stanford, Calif.: Stanford University, Computer Science Department, 1973.

Varian MAT GmbH.: *Operating manual for mass spectrometer MAT 711*. Bremen: Varian MAT GmbH., 1971.

Varkony, T. H., R. E. Carhart, and D. H. Smith. Applications of artificial intelligence for chemical inference, XXIII. Computer-assisted structure elucidation. Modelling chemical reaction sequences used in molecular structure problems. To appear in W. T. Wipke (Ed.), *Computer-assisted organic synthesis*. Washington, D. C.: American Chemical Society, 1976.

Varkony, T. H., R. E. Carhart, and D. H. Smith. Computer assisted structure elucidation, ranking of candidate structures, based on comparison between predicted and observed mass spectra. In *Proceedings of the Twenty-Fifth Annual Conference on Mass Spectrometry and Allied Topics*, Washington, D. C., 1977.

Varkony, T. H., R. E. Carhart, D. H. Smith, and C. Djerassi. Computer-assisted simulation of chemical reaction sequences: Applications to problems of structure elucidation. *Journal of Chemical Information and Computer Sciences* 1978(a), 18, 168.

Varkony, T., D. Smith, and C. Djerassi. Computer-assisted structure manipulation: Studies in the biosynthesis of natural products. *Tetrahedron*, 1978(b), 34, 841.

Waller, G. R. (Ed.). *Biochemical applications of mass spectrometry*. New York: Wiley-Interscience, 1972.

Wegmann, A. Variations in mass spectral fragmentation produced by active sites in a mass spectrometer source. *Analytical Chemistry*, 1978, 50, No. 6, 830–832.

Whewell, W. *Novum organum renovtum* (3d ed.). London: John Parker & Son, 1858.

Wiener, N. *God and golem, inc.* Cambridge: M. I. T. Press, 1964.

Woodruff, H. B., and M. E. Munk. Computer-assisted interpretation of infrared spectra. *Analytica Chimica Acta*, 1977, 95, 13–23.

NAME INDEX

Achenbach, M., xii
Adlercreutz, H., 174, 185
Anderson, P., 179, 182

Bacon, F., 160–161
Banerji, R., 171
Barenfeld, M., 154, 185
Belzer, J., 178, 180
Beynon, J., 22, 179
Boden, M., 157, 179
Braekman, J., 177, 181
Bronowski, J., 162n., 179
Brown, H., xi, xii, 40, 48, 175, 177, 179, 181
Buchanan, B., 49, 51n., 68, 108, 125, 131–133, 144–145, 170–185
Buchs, A., xii, 68, 172–173, 180–181, 185

Caprioli, R., 179
Carhart, R., xi, xii, 14, 24, 101, 108, 134–135, 175–179, 181–186
Carrington, R., 176, 184
Cheer, C., 134, 177, 181
Chomsky, N., 164
Churchman, C., 171, 181
Condillac, E., 161
Creary, L., xii

Daloze, D., 177, 181
Davis, R., 108, 181
Delfino, A., xii, 172–173, 180–181, 185
Djerassi, C., x, xii, 24, 131, 133, 135, 171–175, 177–181, 183–186
Dreyfus, H., 174, 179
Dromey, R., xi, xii, 24n., 70, 142–143, 176–178, 181
Duffield, A., xii, 171–174, 177–178, 180–181, 183, 185
Dunham, L., xii

Eggert, H., xii
Engelmore, R., xii, 174, 185
Erman, L., 159, 181, 184

Ernst, G., 29, 154, 182
Everhart, E., 179, 182

Fateman, R., 127, 154, 183
Feigenbaum, E., 68, 108, 132, 170–175, 177–178, 180–183, 185
Feldman, R., 176, 182, 185
Fisher, F., xii
Fitch, W., 179, 182
Fowler, W., 175, 186
Freiman, C., 173, 180
Friedberg, R., 182

Gaschnig, J., 157, 182
Gelernter, H., 175, 186
Giere, R., 182
Gray, N., xii, 104, 178, 182
Green, C., 38, 182
Gritter, R., xii, 177, 180
Gross, M., 179, 185

Hammerum, S., xii
Hart, A., 175, 186
Hartley, D., 163
Hayes, J., 179–180
Hayes-Roth, F., 179–180
Hearn, A., 127, 182
Heller, S., 26, 176, 182, 185
Hjelmeland, L., xii, 40, 48, 175–176, 179
Hooke, R., 161
Hyde, E., 176, 185

Jennings, K., 182
Johnson, C., Jr., 178, 180
Johnson, S., xii, 176, 181
Jordan, C., 49
Jurs, P., 178, 185

Kibens, M., xi
Kinariwala, B., 170, 182

NAME INDEX

Kleinmuntz, B., 170, 183
Konopelski, J., xii, 135, 177, 185
Kuhn, T., 161, 166, 182
Kuo, F., 170, 182

Laudan, L., 161, 163, 182
Lavanchy, A., xii
Lederberg, J., 2, 19, 40, 127, 129, 131–133, 167, 169–174, 176–178, 180–183, 185
LeSage, G., 161, 163
Lesser, V., 159, 181
Lindsay, R., 14, 155, 183
Ludena, E., 178, 180
Lykos, P., 181

McCarthy, J., 39, 183
McLafferty, F., 16n., 26, 184
Martin, W., 127, 154, 183
Masinter, L., xii, 40, 43, 46, 48, 175–177, 179, 183, 185
Medawar, P., 160, 184
Meltzer, B., 170, 172–173, 180, 182
Mesarovic, M., 171
Michie, D., 170, 172–173, 176, 179–180, 182, 184
Mikulich, L., 179–180
Milne, G., 23, 51n., 173, 180, 182, 184
Mitchell, T., xii, 118n., 178–180, 184–185
Morrell, A., 171, 182
Morrill, K., xii
Mortimer, C., 184
Munk, M., 25n., 186

Neely, R., 184
Newell, A., 29, 87, 154, 182, 184
Newton, I., 161
Nilsson, N., xi, 30, 184
Nourse, J., xii, 5n., 178–179, 184

Peirce, C., 182
Petrick, S., 182–183
Polya, G., 163, 184
Popper, K., 161, 184

Quayle, A., 173

Raphael, B., 154, 184
Reddy, R., 184
Rindfleisch, T., xii, 177–179, 181, 184, 186
Robertson, A., xii, 49, 51n., 171, 173, 180–181, 183
Robinson, J., 29, 153, 184
Rouvray, D., 127, 185

Sabelli, N., 178, 180
Schroll, G., xii, 171–172, 180–181, 185
Schwenzer, G., xii, 178, 184–185
Sheikh, Y., xii, 133, 172, 185
Shortliffe, E., 151, 185
Shue, H., 175, 186
Simon, H., 147, 154, 184–185
Smith, D., xi, xii, 43, 46, 66, 85, 101, 128, 134–135, 137–138, 141, 176–186
Smith, R., 178, 180
Sridharan, N., xii, 43, 46, 173–174, 176–177, 180–181, 183, 185–186
Stefik, M., xii, 177–178, 181
Sutherland, G., xii, 40n., 68, 108, 170–173, 181, 183, 185–186

Teitelman, W., 39, 186
Toulmin, S., 162, 186
Tursch, B., 177, 181

Van Antwerp, C., xii
VanLehn, K., 39, 186
Varkony, T., xii, 66–67, 135, 177–179, 181–182, 186
Venkataraghaven, R., 26, 184

Wahl, A., 178, 180
Waller, G., 104, 174, 178–179, 182–184, 186
Waterman, D., 179–180
Wegmann, A., xii, 179, 186
Westfall, R., 182
Whewell, W., 167, 186
White, W., xii, 40n., 174, 177, 180, 185
Wiener, N., 149, 186
Winograd, T., 157
Wipke, W., 176–177, 185–186
Woodruff, H., 25n., 186

Yeager, W., xii, 178, 186
Yeo, A., xii, 174, 185

SUBJECT INDEX

Absorption spectra, 15
Abstraction, 33
Abstraction planning, 33, 35
Acetaldehyde, 10
Acetals, 129, 143
Acetamide, 10
Acetic acid, 7, 10
Acetone, 10
Acid(s), 7, 129, 142
   acetic, 7, 10
   amino (*see* Amino acids)
   aromatic, 158
   formic, 10
   methanoic, 7, 10, 87
   organic, 134
   urinary organic, 179
Action-part of productions, 87, 89–91, 122
Acyclic compounds, 9
Acyclic DENDRAL algorithm, 48
Acyclic DENDRAL notation, 171
Acyclic generator, 38, 40–41, 48, 53, 69, 106, 127, 170, 173
Acyclic heuristic DENDRAL, 130
Acyclic isomers, 127, 176
Acyclic structure diagram, 49
Acyclic structures, 2, 40, 48, 52, 171
ADRAW, 79
Advanced Research Projects Agency (ARPA), xii, 152, 178
AI (artificial intelligence), 1, 28, 147
Alcohols, 7, 9, 11, 69, 89, 95, 131, 142–143, 158
Aldehydes, 7, 10, 92, 95, 129
Algorithm(s), 32, 151, 156, 159
   acyclic DENDRAL, 48
   Chinese menu, 84
   DENDRAL (*see* DENDRAL algorithm)
   generating, 167
   labeling, 48
   notation, 167
Aliphatic compounds, 2, 9, 49n., 106
   amines, 109, 131, 143, 173
   ethers, 171–172
   ketones, 69, 131
   planning rules for, 68
Aliphatic isomer, 129
Aliphatic structure, 48
Alkanes, 6

Alkenes, 6
Alkyl side chain, 67
Alkyls, 6
Alkynes, 7, 66
ALLBREAKS, 101
ALLBRKS, 38
Amides, 8, 10
Amines, 8, 10–11, 69–70, 93, 95, 133, 143, 158
   methyl, 10
   saturated, 69–70, 172
Amino acids, 8, 11, 69, 101, 130–131, 134
   derivatives, 136, 143
Amino group, 8
Ammonia, 8, 10
Amu (atomic mass units), 16
Androstane, 111, 114–118
   skeleton, 110, 118
Antibiotic therapy, 151
Apical node, 51
Aromatic acids, 158
Aromatic bond, 101, 158
Aromatic ring, 109–110
Aromaticity, 150
ARPA (Advanced Research Projects Agency), xii, 152, 178
Artificial intelligence (AI), 1, 28, 147
ATNAME, 57–59, 61
Atom:
   primary, 47
   quaternary, 47
   secondary, 47
   TAGed, 54
   tertiary, 47
   univalent, 11, 44
   vertex, 41
Atom type, 119
ATOMFV command, 56
Atomic mass units (Amu), 16
Atomic number, 52
Automatic programming, 159n.
AZOCINE, 65

Bad compositions, 72–73
Bad losses, 72–73, 75
BADLIST, 38, 84, 126, 129–133, 165, 170–171

189

190 SUBJECT INDEX

Ball-and-stick model, 4, 27
Barbiturates, 136
BCPL, 39
B/E (break environment), 118–120
B/E set, 119, 122
Benzene, 12
Biosynthetic constraint, 67
Bit, 156n.
Body fluids, 134
Bonding sites, 4
Bonds, 4
 alpha, beta, gamma, 13n.
 equivalent, 7
 single, double, triple, 4
Bottom-up exploration, 165
BRANCH, 79–80
Breadth-first search, 148
Break analysis, 137
Break environment (B/E), 118–119
Break function, 90
BREAKS, 82–83
Bredt's rule, 158
Butane, 6, 52
Butanone, 10
Butyl radical, 52

Canons of order, acyclic DENDRAL, 49–50, 53
Carbon black adsorbates, 179
Carbon dioxide elimination, 158
Carbon monoxide elimination, 158
Carbonyl group, 7, 47, 55, 103
Carboxyl radical, 7
CATALOG, 42, 47, 177
Centroid, 49–51, 53
CHAIN, 56, 58–59, 61, 79
Chemical graph, 5, 49, 149, 155, 157
Chemical nomenclature, 48
Chemical reactions, 66
Chemical stability, 126, 130n., 133n., 158
Chemical structure notation, 170
Chess, 31, 34, 39, 108, 154
Chinese menu algorithm, 84
Chlorinated hydrocarbons, 177
Chromatography, 24
Ciliated skeleton, 46, 48
Class-specific fragmentation rules, 103, 105, 144
CLEANUP, 24n.
Cleavage, 13
 alpha, 13n., 70, 104, 143
 beta, 13n., 104, 143
 gamma, 13n., 91, 143
Cluster, 71
Coarse search, 119, 124–125
Completeness, 32, 41, 165
Compounds:
 acyclic, 9
 aliphatic (see Aliphatic compounds)
 biologically active, 6
 cyclic, 9
 drug, 143
 mixtures of, 24n., 136, 139
 saturated acyclic monofunctional, 69
Computer technology, 29, 168
Condenser, deflection, 15, 20
Configuration symmetry group, 178
CONGEN (constrained generator), 2, 37–38, 41, 53–57, 64–65, 67–68, 79, 85, 100–101, 103, 105–106, 127, 133–135, 144, 147–150, 168, 177
Conjugation, 55, 58

Connectivity, 27
Connectivity isomer, 4, 6, 41, 49, 54, 64, 107
CONSTRAINTS, 57
 ISOPRENE, 54
 LOOP, 60
 PROTON, 54–55, 60
 RING, 54, 60–62
 SUBSTRUCTURE, 54, 60–62
Constraints, 38, 54, 68, 83–84, 86, 106, 120, 164, 177
 geometric, 4
 heuristic, 177
 topological, 4
Crystal structure, 159
Custom crafting, 107
Cyclic compounds, 9
Cyclic isomers, 176
Cyclic ketones, 9–10
Cyclic skeleton, 46–48
Cyclic structure generator, 49n., 53, 68, 151, 172, 175–177
 (See also CONGEN)
Cyclohexanone, 10

Data-driven exploration, 165
Data-driven heuristics, 160
Data-driven planning, 159
Daughter ion, 77
Daughter peak, 87
DB, 90
Decane, 52
DEFINE, 54, 56–59, 61
Degree, 47
 of unsaturation, 44, 158
Degree list, 46–47
DENDRAL algorithm, 169
 notation, 48, 169, 173
 PLANNER, 36, 68, 79, 105
 planning, 86
DENDRAL's knowledge, 157
Depth-first search, 148
Descriptive science, 162
Design principles, 126, 147
Determinism, 29
Diallylic proton, 55, 58
Dialogue, 56–57, 81, 108
Digital Equipment Corporation, 24n., 39
Diols, 129
Dipole moment, 159
Discovery, 2, 40, 68, 86, 165
Discrimination, 87, 100, 102–103, 146
Dominant series, 73
DONE, 58, 61
Dots, 88, 119
DRAW ATNAMED, 57–63
DRAW NUMBERED, 57
DRAWSOME, 57
Drift region, 15, 20–22
Drug compounds, 143
Dynamic stereochemistry, 178

Electric field, 13
Electron beam, 13
Electronegativity, 159
Electrons, shared, 4
Electrostatic section, 20
Empirical formula, 4, 11, 42, 44, 49, 54, 69, 78, 106, 127, 158
Empirical support, 84
Enumeration, 2, 9, 36, 40–42, 49, 51, 162
 exhaustive, 161
Esters, 7, 129, 142–143

SUBJECT INDEX 191

Estradiol, 84, 98–100, 138
Estriol, 139
Estrogen breaks, 81
Estrogen skeleton, 11, 81, 143
Estrogenic steroids, 9, 11, 77, 79, 81, 83–85, 90, 93, 95, 133, 135, 137, 142–143, 146, 158, 173–175
Estrone, 138–139
    16 alpha-hydroxy-, 135–137
Etamide, 10
Ethanal, 10
Ethane, 6, 52
Ethers, 7, 9, 11, 69, 92, 95, 129, 131, 143, 158
Ethyl radical, 6, 52
Evaluator, 172–173, 176
Evidence, 19, 78, 84, 116–118, 122, 138n.
    negative, 123, 144, 145n.
    positive, 145n.
Evidential support, 77, 84
EXAMINE, 64–65
Exhaustive exploration, 32
Exhaustive search, 40, 113
Expectation-driven exploration, 165
Expectation-driven heuristics, 160
Expectation-driven planning, 159
Explanation, 31, 35, 85, 101, 106, 113–115, 118, 120, 131n.–133n., 162, 166
Explanatory science, 162
Exploration:
    data-driven, 165
    heuristic, 155, 164
    random, 32
    selective, 33
    systematic exhaustive, 32

Feedback criteria, 84n.
Feedback loop, 135, 138n.
Formaldehyde, 10
Formic acid, 10
FORTRAN, 149
Fragments, 13, 20
Free valence, 79
Free valence partition, 48
Functional groups, 7, 25, 65, 129
Fused-ring system, 101
Fused rings, 158

Game-playing programs, 154
Gas chromatography, 24
Gas chromatography/mass spectrometry (GC/MS), 24, 134, 177–179
GC/MS (gas chromatography/mass spectrometry), 24, 134, 177–179
General Problem Solver (GPS), 29, 38, 154, 163
Generalization, 118, 120, 166
Generalized break analysis, 78
Generalizing, RULEMOD, 124
GENERATE, 55–57, 60
Generate and test, 34, 36, 68
Generating algorithm, 167
Generation, 33, 76, 106, 149, 159
    ring, 44
    stereoisomer, 64, 178
GENERATOR, 170
Generator, 36, 40, 84n., 148
    acyclic, 38, 40–41, 48, 53, 69, 106, 127, 170, 173
    cyclic, 40, 127
    of plausible solutions, 160
Geometric constraints, 4
Geometric distortions, 178

Goal state, 31
GOODLIST, 38, 68, 70, 84, 126, 149, 170–171
GPS (General Problem Solver), 29, 38, 154, 163
Graph, 2, 5
    labeling, 175
Graph theory, 64n.
Guidance methods, 159

H-transfer, 103
    (See also Hydrogen transfer)
Half-order theory of mass spectrometry, 100–103, 105, 109, 115, 144, 158
Hamilton circuit, 169–170
HEARSAY, 159
Heptane, 52
Heteroatom, 7, 52, 69–70, 130
Heuristic, 1, 53, 147, 157
Heuristic constraints, 177
Heuristic DENDRAL, 35, 38, 105
Heuristic exploration, 155, 164
Heuristic generation, 34
Heuristic method, 32
Heuristic programming, 1, 28, 127, 151, 171
Heuristic programs, 154, 159
Heuristic search, 33, 53, 167–168, 175
Heuristic theory formation, 173
Hexane, 51–52
High-resolution spectrum, 20, 70, 132
Hill climbing, 35, 149, 159, 163
Hosohedron, 42–43, 47
HRANGE, 56–57, 59, 61
Human engineering, 56–57, 65, 79, 81, 134, 148, 150
Hydrocarbons, 6–7, 50
    chlorinated, 177
    saturated, 11–12
    unsaturated, 135
Hydrogen transfer, 22, 73, 78, 81–82, 87, 90, 101, 103, 109, 111, 122, 143, 168
Hydroxyl radical, 7
Hypothesis formation, 28, 36, 164, 167–168
Hypothesis generating, 151
Hypothesis testing, 86, 151
Hypothetico-deductive scheme, 160

IMBED, 55–57, 61–62
Imbedded superatom, 61
Imbedding, 42, 60
Induction, 125, 160, 171
    by elimination, 161
Induction program, 107
Inductive inference, 162
Initial partition, 44
Initial state, 31
Inlet system, 16
Insect secretions, 135
Intensity function, 90
Interdisciplinary research, 1, 151–152
INTERLISP, 39, 87–89, 94, 149
International Joint Conference on Artificial Intelligence, 154
INTSUM, 38, 103, 109–110, 113, 115–116, 119, 144, 156n., 165, 174–175
    table of information, 115
Ion(s), 13
    daughter, 21
    molecular (see Molecular ions)
Ion list, 87
Ion structures, 135
Ionization chamber, 16
Ionization voltage, low, 77

192 SUBJECT INDEX

ISIT, 90
Isocane, 52
Isomer(s), 4, 158
 acyclic, 127, 176
 aliphatic, 129
 connectivity, 4, 6, 41, 49, 54, 64, 107
 cyclic, 176
 number of, 52, 70, 105, 127–129, 131–132
Isomerism, 4
 structural, 177
  scope of, 127
Isoprene rule, 135, 158
Isoprene unit, 54, 65
Isotopes, 16, 19, 71, 77

Jigsaw puzzles, 35
JOIN, 61, 79–80

Keto group, 7, 55, 57, 102, 110, 114
Ketoandrostanes, 122, 143, 146, 158
 di-, 125, 144, 146
 mono-, 100, 102, 110, 113, 116, 121–123, 125, 143, 145–146
 tri-, 125, 144, 146
Ketones, 7, 10–11, 89, 92, 95, 129, 131–132, 143, 158, 171
 cyclic, 9–10
 isomeric, 172
 methyl ethyl, 10
 mono-, 146
"Knit," 156n.
"Knowing how," 156
"Knowing that," 156
Knowledge:
 informal, 2, 55, 85, 151, 165, 168
 multiple sources of, 34
 procedural imbedding of, 94
 task-specific, 1, 30, 36, 38–39, 127, 153–154, 166
 uniform representation of, 151
Knowledge acquisition, 39, 107–108, 125, 172
Knowledge-based systems, 36, 148, 154, 176
Knowledge engineering, 39, 153, 155
Knowledge representation scheme, 155

Labeling algorithm, 48
Laboratory notebook, 149–150
Learning, 2, 171
Learning program, 125
Library search, 26–27
LINK, 56–57
LISP, 39, 49n., 88n., 159, 172
Logic of discovery, 167
Low-ionizing voltage spectra, 174
Low-resolution spectrum, 70, 132

M⁺ (see Molecular ions)
MCLAFFERTY, 94
McLafferty rearrangement, 22–23, 91, 103–104, 158
Macrolide antibiotics, 100, 103, 144
Macrolide skeleton, 103
Macrolide structures, mass spectra for, 104
Macrolides, standard, 104–105
MACSYMA, 127, 154
Magnetic section, 21
Mars probe, 130
Mass:
 accurate, 19
 nominal, 16
Mass spectrometer, 3, 13
 resolution of, 20n.

Mass spectrometry (MS), 1, 3–5, 11, 13, 35, 69, 158
 half-order theory of, 100–103, 105, 109, 115, 144, 158
 zero-order theory of, 69, 101, 109, 118, 130–131
Mass spectrum, 16–17, 35, 38
Mass-to-charge ratio, 15, 20
Mathematics, 31, 127, 154
M/e (see Mass-to-charge ratio)
Melting point, 159
Merging, RULEMOD, 124
Meta-DENDRAL, 107, 145, 150
 results, 143
Metabolic disorders, 134
Metabolic products of microorganisms, 135
Metastable defocusing, 22, 87
Metastable ion, 21, 77, 174
Metastable peak, 22, 87, 99–100, 158
Methanal, 10
Methane, 6, 18, 52
Methanoic acid, 7, 10
Methanol, 7, 9
Method of construction, 151
Method of successive approximation, 163
Methyl radical, 6, 52, 55, 59, 70, 132
Methylpropane, 6
Methynolide, 104–105
Microgram, 16
Milligram, 16
MISSING, 90
Mixtures of compounds, 24n., 136, 139
Molecular ions, 16, 38, 69–71, 73–74, 76, 87, 136, 150, 177
MOLION, 38, 69–70, 73–74, 77–78, 85, 106, 113, 136, 142, 151, 165, 177
 postulate, 71–72
Mono-hydroxy steroids, 178
Mono-terpenoid skeletons, 177
Monofunctionals, saturated, acyclic, 131, 171
MS (mass spectrometry), 13
 high-resolution, 19
 low-resolution, 19
 (See also Mass spectrometry)
MS planning rules, 70
MS process, 109, 113
MSPRUNE, 38, 101
MSRANK, 102–103
MYCIN, 151

NASA (National Aeronautics and Space Administration), xii, 169–170, 172
National Aeronautics and Space Administration (NASA), xii, 169–170, 172
National Institutes of Health (NIH), xii, 152
Natural languages, 29
Natural order, 53
NDRAW, 79–80
Neighbor, 119
Neutral fragment, 113
Neutral losses, 101, 103
Neutral molecule loss, 78
Neutral transfer, 103
NIH (National Institutes of Health), xii, 152
Nitrile, 66
Nitrogen heterocycles, 135
Nitrogen parity, 24, 73
Nitrogen rule, 23–24
NMR (see Nuclear magnetic resonance)
Noisy data, 19, 71, 151
Nominal mass, 17, 19
Nonane, 52

SUBJECT INDEX **193**

Nonheuristic methods, 32
Normal science, 161
Notation, $6n.$, 48, 89
Notation for hypotheses, uniform, 2
Notational algorithm, 167
Nuclear magnetic resonance, 24, 26, 132, 171–172
   ^{13}C, 175, 178
Nuclear magnetic resonance spectrometry, 24–25
   ^{13}C, 107

Octane, 52
ON, 90
Ordering, natural, 51
Organic acids, 134
Organic chemistry, 3–4, 6
Oxime, 93, 95

Partial solutions, 159
Pentane, 52
Pentyl radical, 50–52
Permutation, 41
Peroxides, 129
Pesticide conjugates, 135
Pharmacologic agents, 135
Plan-generate-test, 53, 69, 71, 153, 161, 164
PLANNER, 38, 78, 84–85, 87, 90, 106, 126, 135–137, 139, 150, 170, 173–174, 176
Planning, 33, 68, 75, 105, 133, 148, 159, 166
Planning program, 36–37, 53
Planning Rule Generator, 38, 68–70, 81, 105, 133
Planning rules, 38, 81, 173
   for aliphatic compounds, 68
Polish prefix notation, 49
Polymeric carbons, commercial, 179
Polymeric structures, 150, 159
Poor primary-loss list, 74, 76
Poor secondary-loss list, 74, 76
Positive evidence, 115
Predicate calculus, 29, 38, 154
PREDICTOR, 37–38, 87, 92, 98, 106, 126, 132, 135, 144, 165, 170–173, 176
Predictor and Evaluation function, 170
Preliminary Inference Maker, 132, 170–173
Primary fragment, 72
Primary loss, 71–72, 76
Prism, 43
Problem:
   of discovery, 160
   of verification, 160
Problem solver, 31
Problem solving, 2, 29–30, 153, 155, 164, 171
Problem space, 30, 129, 149, 154, 164
Problem states, 30, 149
Problem-solving paradigm, 30
Problem-solving systems, 38
Problem-state languages, 30, 35
   secondary, 31, 33
Production:
   action-part of, 87, 89–91, 122
   situation-part of, 87, 89–91, 122
Production system, 87
Production-system-oriented program, 106
Productions, 69, 81, 87, 90, 109, 151, 157, 164, 179
   representation of, 89
Programming, 151
   automatic, $159n.$
   heuristic, 1, 28, 127, 151, 171
   languages: list-processing, 29
      string-processing, 29

Proof by eliminative induction, 161
Propane, 6, 49, 52
Propanone, 10
Propyl radical, 6, 51–52
Protein molecules, 9
Proteins, 8
Proton NMR planning rules, 70
Proton NMR spectrum, 69
Proton nuclear magnetic resonance, 70, 173
Pruning, 42, 160
Pruning methods, 159
"Purl," $156n.$

Question-answering systems, 38

R, 5
Radical, 6, 51
Random generation of hypotheses, 163
Ranking, 70, 74, 86–87, 100, 102–103, $131n.$, 142, 144, 146–147, 178
REACT, 64, 66–67
Reaction, chemical, 66
Rearrangement, 22, 158
   McLafferty, 22–23, 91, 103–104, 158
Rearrangement products, 135
REDUCE, 127
Refined search, 124–125
Relative abundance, 16, 90
Remainingpot, 42, 44, 48, 53
Representation, 37, 88
   uniform, 37
Representation language, 5
Representation problem, 31, 35
Resolution, 29, 37, 153
Resonance, 7
RING, 56–58
Ring-free structures, 2
Ring fusion, 65
Ring generation, 44
Ring structures, 2
Ring superatom, 41–42, 44, 47–$49n.$, 53–54
Ring superatompot, 42, 47, 53
Rings, 12
   fused, 41
      edge-, 12, 41, 101
      spiro-, 41–42
Robotics, 154
Rough approximation, 125
Rule formation, 173
RULEGEN, 38, 109, 115, 118–119, 121, 125, 165
   templates, 123
RULEMOD, 38, 109, 123

SAIL, 39
SAM (saturated acyclic monofunctional) compounds, 69
Saturated acyclic monofunctional (SAM) compounds, 69
Saturated amines, 69–70, 172
Saturated hydrocarbon, 11–12
Scanning electron multiplier, 21
Scientific discovery, 68, 153, 160
   model of, 164
Scientific inference, 171
Scientific paradigm, 161, 165, 167
Scientific reasoning, 153
Score, 146
Scoring function, 102–103, 122, 144, 171–172
Search, 33, 107, 149, 159
   heuristic, 33, 53, 167–168, 175
   state-space, 35

Search space, 127
  size of, 131n.–133n.
Secondary fragment, 71–72
Secondary loss, 71–73
Secondary-loss candidate, 75
Selection, RULEMOD, 124
Sesquiterpenoid skeletons, 177
SHOW, 57–58
SHRDLU, 157
Situation part of productions, 87, 89–91, 122
Size of molecules, limit on, 11
Skeleton, 109
  androstane, 110, 118
  ciliated, 46, 48
  cyclic, 46–48
  estrogen, 11, 81, 143
  macrolide, 103
  mono-terpenoid, 177
  sesquiterpenoid, 177
  sterol, 67
Solution space, 159
Specialization, 166
  RULEMOD, 124
Spectra(um):
  high-resolution, 20, 70, 132
  low-ionizing voltage, 174
  low-resolution, 20
  mass, 16–17, 35, 38
  proton NMR, 69
Spectrometer:
  double-focusing, 20–21
  mass, 3, 13
    resolution of, 20n.
Spectrometry:
  infrared, 24–25
  mass (see Mass spectrometry)
  nuclear magnetic resonance, 24–25, 107
  ultraviolet, 24–25
Spectrum list, 87
Spiro forms, 65, 158
Spiro-fusion, 47
Stanford University, 27, 39, 53n., 65n., 131, 151
STEREO, 64
Stereocenter, 64–65
Stereochemistry, dynamic, 178
Stereoisomer generation, 64, 178
Stereoisomerism, 5, 158, 178
Stereoisomers, 4, 64–65, 110, 158, 179
Steroid nucleus, 109
Steroids, 11, 102, 110
  estrogenic (see Estrogenic steroids)
Sterol skeleton, 67
Sterols, 9, 67
  marine, 11, 67, 135, 147, 158
STRUCC manual, 65n.
Structural isomerism, 177
  scope of, 127
Structure editing, 56, 65, 79
Structure elucidation, 1, 3–4, 19, 27–28, 35, 144, 159
Structure Generator, 170–173, 176
Structures:
  acyclic, 2, 40, 48, 52, 171
  aliphatic, 48
  chemical, teletype printing of, 177
  crystal, 159
  ion, 135
  macrolide, mass spectra for, 104
  polymeric, 150, 159
  ring, 2

Structures:
  ring-free, 2
  topological, 64
Subproblem space, 159
Substituents, 5, 69, 103
Sulfhydryl group, 8
SUMEX-AIM, 39, 176–177
Superatom, 38, 54–57, 61, 134–135
  imbedded, 61
  ring, 41–42, 44, 47–49n., 53–54
Symmetry, 41, 47–48, 158
Synthesis program for complex organic molecules, 175

TAG, 59
TAGged atoms, 54
Task-directed method, 30
Task-specific information, 154
Tautomerism, 158
TEIRESIAS, 108
Teletype printing of chemical structures, 177
Template, 119–120, 122
Template refinement, 121
Terpene rule, 158
Terpenoids, 135
TESTER, 87, 172
Testing, 36–37, 53, 74, 76, 101, 106, 133, 148
Tetrahedron, 43
Thioesters, 8
Thioethers, 8, 69, 92, 95, 131, 143, 158
Thiols, 8, 11, 69, 92, 95, 131, 158
Top-down exploration, 165
Topological constraints, 4
Topological structure, 64
Topological symmetry, 178
Topology, 4–5, 18
Transfer function, 90
Transformation, symmetry, 53
Transformations, 30
Trees, 9
  generation of, 41–42, 48
Trifluoroacetonitrile, 18

Uniform proof procedure, 29, 37–38
Uniformity of representation, 150, 155
Univalent atoms, 11, 44
Unsaturated hydrocarbons, 135
Unsaturation, 12, 52, 78, 88, 127–128
  degree of, 6, 11–13
Unseparated mixture, 174
Urinary organic acids, 179
Urine, 24n.

Valence, 2, 4, 156
  free, 41, 44–45, 53
Valence list, 44
Varian MAT, 15, 186
Verification, 86, 160
Version spaces, 179
Vertex atom, 41
Vertex graphs, 41–43, 46–47, 171, 177
Viking mission, 130
Vinyl group, 55, 58
Vinyl proton, 55

Water, 4, 17–18, 78
Water elimination, 158
Working backward, 34

Zero-order theory of mass spectrometry, 69, 101, 109, 118, 130–131